INTERDIMENSIONAL

Consciousness, Cryptids and the Hidden Architecture of Reality

Ronny LeBlanc

This book is a work of nonfiction. Certain names, identifying details, and chronological elements have been altered for clarity and privacy while preserving the factual and thematic integrity of the events described.

The views and interpretations expressed in this book are those of the author and are presented for informational and exploratory purposes. They do not constitute scientific, medical, or legal advice.

First edition

Printed in the United States of America

ISBN: 979-8-218-90537-8

TABLE OF CONTENTS

Chapter One

THE THUNDERBIRDS & TELEPATHY

In recent years, there have been reports of sightings that defy conventional explanation—strange lights in the sky, mysterious creatures and encounters with beings that seem to deny the laws of physics. My wife and I had one ourselves.

Two weeks before my family and I moved to our new home, my wife Amy and I had an amazing experience. We witnessed several UFOs or *UAPs* one after the other. It was late June and we were scheduled to close and move into our new home on July 3rd, 2019. Our house of eighteen years didn't disappoint. We had experiences inside and outside the house that leaned on the side of the supernatural or the paranormal.

We were sitting in our two-car garage with the door open, the sun had settled and the sound of crickets and summer echoed as we talked. We were sitting on a couple of bar stools with my back facing the open garage door and my wife Amy, wearing shorts, in front of me.

I have shared this story over a hundred times during my lectures, library tours and conferences around the country and it still blows me away. When we vocalize something out loud and speak it forth, something happens, someone or something hears it. Be careful what you wish for. They, whoever they are, just might respond.

I was excited. "I can't wait to move to the new house. We are right down the road from Leominster State Forest (aka Monsterland)...we are going to see some crazy stuff!" My wife agreed as we had already experienced things in Leominster that defied explanation. As we began to discuss logistics and what we would need to start packing and deciding on what items would not make the trip, Amy's eyes widened without warning. I could immediately see fear in her eyes.

"Ronny, something just touched my leg." I looked at her, assessed the situation quickly and surmised that maybe it was just a moth or a Japanese beetle that may have grazed or touched her leg and flew away coming in from the open garage door. "Maybe it was just a bug?"

She shook her head. "No, this feels as if someone is putting a hand on my leg." We both paused and after a few moments with a little apprehension we started where we left off with the conversation.

In what seemed roughly not even a minute, it happened again. Amy frantically called it out. She was visibly shaken. This was getting weird. As soon as we were talking again it happened for a third and final time. Amy immediately shot out of her seat and exclaimed, "Okay, I am freaking out. It feels like feathers just grazed my leg this time. I am getting strange fairy vibes!"

I then jumped off the stool with the immediate realization that someone or something was trying to get her attention. But why?

"Maybe we should go outside?! It feels as if someone is trying to get our attention", I said. We scuttled out of the garage and onto the driveway, a few yards away the front porch and entryway to the house was visible and the timing of what we were about to see was somewhat divine. It was perfectly timed. Our jaws dropped as we looked above the roof of the house and to the left we saw something high above but lower than a plane. I have never seen anything like this before and I hope I DO get to see it again.

We watched as a "thunderbird" made of red flames and orange plasma, the size of a school bus soared silently in the sky, swimming across as it playfully chased another thunderbird made of green flames and a plasma-like substance. They were the same size and same shape. Amy and I both agreed they were the same type of being, just different colors. Another description of what we witnessed could be explained as a fiery phoenix or even a dragon. Supposed mythical creatures right?

These thunderbirds resembled dolphins in the way they moved, undulating up and down as if under the ocean and immediately the saying "As Above So Below" flooded into my mind. We watched, stunned as these enormous, very alive and real creatures in the sky entertained us with wonder and amazement until they flew out of view. "What the fuck was that!? Are you kidding me? Are we on the set of a science fiction film?!" I asked my wife.

Aside from several "F bombs" and gasps, we couldn't believe what we just witnessed. We were stunned and in a complete state of shock. I was excited. Another thought popped into my mind like the voice in your head as you are reading this book."*Don't go anywhere. There are more coming.*" I immediately shared this thought with Amy and we began to scan the sky looking for other anomalies. It didn't take long.

Maybe two minutes later we see a beach ball-sized glowing green orb off in the distance and heading our way. This time, a closer proximity to our house and almost directly over the roof. We anticipated its arrival and I realized I was holding my breath as we watched its approach. As it got closer I saw something spinning inside of the orb! In the Bible, specifically in the book of Ezekiel, it talks of a vision where Ezekiel witnesses four creatures inside a wheel within a wheel. Some scholars believe it was a "throne chariot" or Merkabah vehicle that is found in Jewish mysticism. It is considered to be a craft of the higher beings that God created.

I focused on this spinning gyroscopic looking mechanism inside this orb. The orb was just a few feet above our roof line and as it passed by we watched intently as this orb flared up so bright...as if to acknowledge that it knew *we were watching it.* Silent, it followed the thunderbirds path and continued on. We didn't say a word. We looked at each other in amazement. This entire time I didn't even think about pulling out my phone once to record what we were witnessing. Quite frankly, I was afraid to look away.

Amy however, spotted something out in the distance heading in our direction. It was white and blue and moving in the same manner, up and down. "Is that another one honey!? Right there! It's coming this way!" Swiftly and without hesitation, she pulls out her phone and begins to record its entry into our view. It moved like a sparrow, dipping down to the point where it looked as if it was going to fall out of the sky and at the last second, fluttered upwards. "It looks like a freaking bird!", I said.

Amy was able to capture about twenty two seconds of this entity, which resembled a smaller version of the previous thunderbirds that flew directly overhead. We spun our bodies around and continued to follow it with our gaze. It did not make a sound and we watched as it cleared the treeline and disappeared. After we reviewed the video several times once we got inside we noticed something extraordinary. If you freeze the video right before it disappears beyond the trees...you see wings.

Did we witness angels, aliens, creatures from another dimension? Or something natural that has been here for ages, maybe even before humans? Have they been living alongside us all this time? Are they entering our reality from somewhere else...another dimension that could be right next door where they enter and exit our reality at will?

That night forced me to confront a possibility that I have long suspected: that everything that follows in this book: UFOs, thunderbirds, cryptids and telepathic encounters are not separate anomalies, but expressions of a single, layered intelligence interacting with human perception across time, culture and the environment.

Jacque Vallee has long argued that the phenomenon behaves less like visitors from another planet and more like an intelligence capable of manipulating physical reality itself.[1]

Interdimensional Theory: A New Perspective

The concept of interdimensional travel posits that beings from parallel dimensions can traverse into our reality. These dimensions, according to theoretical physics, exist alongside our own, with different physical laws and realities. The notion of a "multiverse" suggests that these dimensions are not just theoretical but could be physically accessible under certain conditions.

UFO sightings, Bigfoot encounters, and paranormal experiences are usually treated as separate mysteries. Yet they often appear in the same locations, during the same periods of instability, and under similar environmental conditions. *Interdimensional* asks whether the issue is not the phenomena but the way we separate them.

Let's start with the fact that many UFO encounters involve phenomena that defy the known laws of physics, such as instant accelerations, right-angle turns at high speed, and the ability to appear and disappear instantaneously. These characteristics align more closely with interdimensional travel than with conventional extraterrestrial spacecraft. Numerous reports describe UFOs vanishing or disappearing in a manner inconsistent with simple travel into space. Witnesses often describe these objects as "fading away" or "slipping into another dimension." As Jacques Vallee observed, "The UFO phenomenon represents a technology that manipulates physical reality by means we do not yet understand."[2]

Encounters with entities or beings described as "alien" often include behaviors and abilities suggesting they are not bound by our physical laws. Examples include telepathic communication, shape-shifting, and manipulation of time and space. You also have reports of **orbs and multicolored lights** that also simply appear and disappear without a trace. These phenomena are often reported in conjunction with the appearance of larger craft, suggesting a complex, multi-layered nature of interdimensional activity.

The paranormal makes an appearance before, during and after these UAP events known commonly now as "the Hitchhiker Effect'. UFO experiences can sometimes be accompanied by poltergeist-like activity, such as objects moving on their own, strange noises, and electrical disturbances. This crossover between UFO phenomena and ghostly activity also hints at a shared interdimensional origin. John Keel noted that these events rarely occur in isolation, writing that "the same areas that produce UFOs also produce monsters, poltergeists, psychic events, and religious visions."[3]

Due to my work on Discovery Channel's Expedition Bigfoot and my research and personal experiences for my Monsterland books I have explored that there is more to Bigfoot and I believe there is a connection to the UFO enigma as well.

Bigfoot: An Interdimensional Entity?

Bigfoot encounters also exhibit characteristics that support the interdimensional hypothesis, despite extensive efforts to track and capture Bigfoot, the creature remains elusive. Witnesses often describe Bigfoot disappearing into thin air or fading into the forest, suggesting the possibility of interdimensional movement.

Many Bigfoot sightings are accompanied by other paranormal phenomena, such as orbs, strange lights, and unexplained sounds affectionately referred to by some as "High

Strangeness”. This clustering of phenomena suggests a connection to broader interdimensional activity.

Where is the physical evidence? Footprints and other physical traces of Bigfoot are often found in remote areas, but these traces rarely lead to a physical encounter. The instant appearance and disappearance of these traces suggest the creature might not be bound to our dimension.

Sasquatch experiencers and witnesses have reported hearing strange vocalizations and infrasound (sound below the range of human hearing) in areas where Bigfoot sightings occur. These sounds are sometimes described as "otherworldly." Some even reluctantly report telepathic or clairvoyant connections with Bigfoot, suggesting a level of consciousness or intelligence that transcends our understanding of physical reality. Keel described the phenomena as adaptive rather than mechanical, stating that “these manifestations seem to behave like a living system, adapting itself to human culture.”[4]

UFO sightings, Bigfoot or cryptid encounters, and paranormal experiences are usually treated as separate mysteries. Yet they often appear in the same locations, during the same periods of instability, and under similar environmental conditions. *Interdimensional* asks whether the issue is not the phenomena but the way we separate them.

After the experience my wife and I had with these plasma-like beings I thought about Thunderbirds simply being a myth or more biological in nature until we saw it for ourselves. In the vast expanse of Native American mythology, the Thunderbird stands as a testament to the power and mystery of the natural world. From the shores of the Pacific Northwest to the plains of the Midwest, stories of this magnificent creature echo through the ages.

Sky Sovereigns: Are They One in the Same?

The Thunderbird is more than just a creature of legend it is also what it represents. It is a symbol of the untamed forces of nature, capable of unleashing both destruction and renewal. In the traditions of the Cherokee and Lakota peoples, the Thunderbird is revered as a divine messenger, carrying prayers and blessings on its mighty wings.

But what is the true nature of the Thunderbird, and how does it fit into the larger tapestry of global mythology? Some scholars believe that the Thunderbird shares a common ancestry with other legendary creatures, such as the Chinese Dragon and the fiery Phoenix. Could these creatures all be manifestations of the same cosmic forces that govern the universe? Are cultural myths real memories?

The Thunderbird's presence in Native American lore is deeply intertwined with the natural world. It is often associated with storms, particularly thunderstorms, which are seen as manifestations of its power. In some traditions, the Thunderbird is believed to reside in the clouds, where it commands the elements and watches over the land below.

Legends of the Thunderbird vary among different tribes. Among the Ojibwe people, it is said to have the ability to create storms by flapping its wings, while the Iroquois believe it to be a protector of the skies, battling evil spirits and ensuring the balance of the world. Despite these variations, the Thunderbird is universally revered as a symbol of strength, power, and protection.

As we journey across the ocean to the ancient lands of China, we encounter another legendary creature that commands the skies, the Dragon. Unlike the Thunderbird, which is often depicted as a bird-like creature, the Dragon is a serpentine being with scales similar to glistening jewels and the power to control the elements.

In Chinese mythology, the Dragon is a symbol of imperial power and celestial authority, revered as a bringer of rain and prosperity. Its appearance is said to herald the arrival of auspicious events, while its fiery breath can cleanse the land of evil spirits and bring about renewal.

But could there be a connection between the Thunderbird of Native American lore and the Dragon of Chinese legend? Some scholars believe that both creatures share a common origin, tracing their roots back to ancient shamanic practices and the worship of celestial beings. Both are associated with the power of the heavens, capable of shaping the world according to their will.

The Dragon holds a prominent place in Chinese culture, appearing in art, literature, and folklore for thousands of years. It is often depicted as a wise and benevolent creature, offering guidance and protection to those who seek its aid.

In the annals of mythology, few creatures are as iconic or as enigmatic as the Phoenix. Born from the ashes of its own demise, the Phoenix is a symbol of rebirth and renewal, its fiery plumage lighting up the sky with the brilliance of a thousand suns.

In ancient Egyptian mythology, the Phoenix was associated with the sun god Ra, rising each morning to herald the dawn of a new day. But its legend extends far beyond the sands of Egypt, with variations found in cultures around the world. From the Greek Phoenix, which burst into flames and was reborn from its own ashes, to the Fenghuang of Chinese legend, a mythical bird of unparalleled beauty and grace.

But what if the Phoenix is more than just a symbol of renewal? What if it is a living embodiment of the cosmic forces that govern the universe, capable of transcending time and space itself? Some believe that the Phoenix, comparable to the Thunderbird and the Dragon, is a manifestation of the divine, a messenger from the realm of spirits.

The legend of the Phoenix is one of death and rebirth, of destruction and renewal. It is a reminder that even in the darkest of times, there is always hope for a new beginning. The Phoenix's fiery plumage symbolizes the transformative power of change, as well as the eternal cycle of life, death, and rebirth.

When you look deeper into these legendary creatures, we begin to see patterns emerge. Patterns that point to a deeper connection between them than mere coincidence. From the Thunderbird's ability to control the elements to the Dragon's mastery of celestial forces, and the Phoenix's ability to defy death itself, each creature embodies a different aspect of the natural world.

But could they all be manifestations of the same underlying phenomenon? Could the Thunderbird, the Dragon, and the Phoenix be different expressions of a single entity, seen through various cultural lenses. A being that exists beyond the confines of time and space?

Some believe that they are all interdimensional travelers, able to move between worlds at will and communicate with beings from other realms. An intelligence that could very well be plasma-based popping in when there is an expected shift in our reality, a warning or marker that we are about to see a transformation. From my personal experiences in the field, my research and looking at the work that other researchers have gathered before, I am more convinced that this is indeed the case.

Thunderbirds, dragons, and the phoenix all occupy the upper realm as regulators of power (storms, fire and renewal). They exhibit control of the elements, lightning and thunder (Thunderbird), fire and stone (Dragon) and lastly fire and rebirth (Phoenix). They all map back to the same mythic function expressed through different cultural lenses and over many generations.

They seem to be associated with cyclical intervention, and they all seem to embody an interdimensional intelligence with each appearing at moments of imbalance, appearing and acting as a corrective force for a limited amount of time, not a constant presence. Is this a modern day sign from a mythical creature letting us know that we are about to experience a shift? Rather than an invasion, Vallee suggested the phenomenon may function as a long term influence on belief systems, calling it "a control system" that shapes human perception across history.[5]

Why do unexplained phenomena keep appearing together? These phenomena aren't the same thing, but they often emerge under the same environmental and perceptual conditions. Rather than offering belief-based explanations, *Interdimensional* examines recurring patterns across environment, consciousness, and human experience—suggesting that our current frameworks may be too fragmented to explain what we're observing.

Chapter Two

THE CLOAKED ENCOUNTER

A few days after moving into the new house I had to leave to film in Alaska for the documentary television series *Expedition Bigfoot* on the Discovery Channel. This was my fourth season being a part of the team. The time away from home was starting to wear on me, my wife, and my kids. We finally had our dream home which abutted another state park with a hiking trail opening at the end of my driveway begging me to enter. Starting my second week of filming in Alaska I received a frantic call from my wife.

While bringing two of our children to a friend's house she was about a quarter of a mile down the road from our house when unexpectedly something or someone ran out of the woods in front of her SUV from right to left and into and up a sharp hillside leading to other houses. She slammed on her brakes because she thought she was about to hit someone crossing the road.

Amy said it was a tall, slender humanoid figure upright on two legs. She could see the outline of the figure but she could see straight through it resembling static electricity on your television screen! It was cloaked, slightly invisible, with a distortion of light or matter inside the body.[7] My wife pulled over to the right side of the road immediately and looked and listened for movement. They heard nothing. They saw nothing. It was gone.

She said it looked more similar to a tall alien than what would be considered a wide-shouldered Bigfoot...it was tall and thin. Whatever it was, it was definitely trying to conceal itself and the timing of running in front of Amy was odd. It resembled experiences I had growing up in Leominster. Others have since approached me with similar accounts with something invisible, cloaked or camouflaged. There are things living alongside us in our woods that don't want to be seen. The right place at the right time almost as if it wanted to be noticed. Some of the aspects that highlight sasquatch being something else include a cloaking or an invisibility capability and a conscious connection that emerges in the form of mindspeak or telepathy.

What Amy saw with her own eyes is very similar to what we captured on *Expedition Bigfoo*t. The EB team had returned to film and research in Washington State for Season 3. While working alongside a river looking for evidence, Dr. Mireya Mayor and I were scanning the ground, looking for footprints or anything that was out of place. After a little while we discovered a strange looking primitive fish trap that was in the middle of the river. It stuck out like a sore thumb. It was a fish weir-a primitive trap made of stones, sticks,or wooden stakes arranged to funnel fish using the river's natural current. It is designed in such a way to capture a fish that would be making its way into the river's natural current and the strength of the current would keep the fish trapped and unable to escape until it would be retrieved by whomever created it. It is said that Sasquatch had learned how to do this from the Native Americans in the area by observing them and their way of capturing fish.

At this same location, there was an enormous sixty foot plus rock ledge on the opposite side of the river that shot straight up. It was essentially a rock wall, with only a few boulders peeking out where the river touched it. The opposite side didn't have much room for foot traffic and we were not about to go across what looked like a small narrow shoreline of rocks to walk on. Upon finding that fish trap, we started to notice what initially we thought was erosion of the top of the cliff, rocks and dirt rolling down into the river below. Rocks would continue falling down and into the river.

It didn't stop. We would occasionally hear the sound of small rocks falling from the top down the cliff and into the river. Oddly, it seemed to only be directly across from us and didn't seem to happen behind us or in front of us as we worked our way down the river. It was strange. It got to a point where it was possible we had someone or something stalking us along the top of the cliff but it was being sloppy about keeping its presence unknown. Or was it trying to let us know we were not alone?

Frustrated, I stopped in my tracks and looked up towards the cliff edge and thought I had seen a tall dark figure, darker than the rest of the trees. As the rocks continued to crumble down into the river I called out to this mysterious entity and said, "Hey! We know you are up there! Just show yourself already!" Dr. Mayor threw me a questionable glance and promptly told me to keep quiet. I rolled my eyes. We continued on when we immediately heard what sounded like an avalanche erupting across from us. It sounded akin to the earth giving way. Debris and larger rocks began to drop and roll down the cliff and into the water below so intensely it got our attention.

Dr. Mireya and I stood at the edge of the river and swung around with our flashlights pointing directly across from all of this commotion. We stood side by side and we were astonished to see a shadow form from the lights of our flashlights walking on two legs from right to left. If you have ever created a shadow puppet as a kid with a flashlight and your hands, this is what we experienced. We couldn't see it with our naked eye, but it walked into our flashlight beams and revealed itself. The silhouette was in the shape of a lurking Bigfoot carefully walking as if it was unaware its location or presence had been unmasked by our flashlights.

What sent a shiver down my spine was the fact that we could see the shadow clearly and I could see hairs jetting out from the outline of the shadow as it walked across in front of us! There was no doubt in my mind that it was a sasquatch and we just captured this on film! But if this was down here with us, did it slide down the cliff and into the water and that's what created all the chaos or was there another one up on top?! It was smaller and I believe it was a juvenile. We learned that they move in pods or family units. When there's one there are probably more.

With this capture when the episode finally aired the network decided to reluctantly edit out me talking to this unknown something up above us. This annoyed me to no end. Why would they do that? Why take that out? The shadow's abrupt appearance and reveal seemed to be a reaction to my action of calling it out and asking it to show itself to us. And damn, it did. Maybe that is why? It would reveal that these things are not animals but more human or based on what they can do, more interdimensional or alien than we understand at the moment.

While the team argued and questioned what the camera team had actually captured, I went to sleep that night knowing that I had been a part of something mind-blowing and now millions of people would be able to witness it for themselves. These beings can do things that defy physics. The footage doesn't do it justice as to what I witnessed with my own eyes. It was epic.

When I finally returned home from our expedition in Alaska, I was feeling off. I didn't feel as though I was myself and I was starting to wonder if I may have brought something back with me. This is also known as "the Hitchhiker Effect"[6] which has been outlined on History Channel's *Secrets of Skinwalker Ranch* series. Although gorgeous and untamed, Alaska is beautiful but that place is trying to kill you. I couldn't wait to be back home. It was not unlike being on another planet with various unknown threats. I was relieved to be back in

Leominster, Massachusetts. It was weird to be coming back to a new home that I hadn't slept in because of the long filming schedule. Everyone had acclimated to the house but its aura was foreign to me. I was feeling very much like an outsider.

After almost a month the weirdness would resume around the property. I still couldn't seem to shake this feeling. In the month of August, Amy was outside on the deck later in the day basking in the sun and on the phone with her sister. Amy began to hear something large and loud thrashing in the backyard beyond the chainlink fence and by the brook that runs along the back of our property. She started screaming my name to come outside as she expected to see a bear or a moose. Whatever this was, it was big and it seemed to be coming closer. She peered into the woods but couldn't see where the disturbance was coming from or what was creating it. She continued to call for me as the sounds got closer and louder.

Moments later a rock is thrown from the area that the sound was emanating from. The rock barrels towards our metal shed and the impact was so loud that her sister could hear it on the other end of the phone. "BOOM!" I run outside to see what all the commotion was about as my kids are telling me Mom is yelling for me. "Ronny! Someone or something just threw a rock at the shed from the woods. I can't see anything. There is no one there!" I head to the fence line and start scanning around. It is swampy and not easily traversable for someone to remain hidden and attempt to leave without being seen. I sensed something was definitely back there but I couldn't identify it.

THE CONFIRMATION

Sasquatch are known to throw rocks at people that get too close to them or they will use rocks or sticks as a way to let you know to stay put and not get any closer.[8] Witnesses that have experienced this proclaim that the rocks hit with such precision that they seem to land near them or right next to them...but they never hit them. I know that they are around here in central Massachusetts based on my personal experiences and other sightings. But did one just throw a rock at my shed? Is it trying to get our attention? An idea pops into my head. *Remote viewing.*

Chris Ducan, a space operations officer with an extensive military and intelligence background is a close friend and someone that we had on *Expedition Bigfoot* to help us with our search. The first time around, Chris brought in an old World War II technology known as an electromagnetic loop antenna. Every living being gives off a unique energy signature. This instrument provides a way to see through the dark and if we are lucky, it can pick up on something comparable to a Bigfoot which would give off a potential primitive or human-esque signature.

As far as we knew - this hasn't been used before in this particular instance in the search for sasquatch. We used the technology in Kentucky and we generated some thought-provoking results. The second time we had Chris on the show was strictly based on my request to bring him back as I knew that he was a "Picasso" when it came to the art of Remote Viewing.[9] RV has also garnered its military moniker "Psionics" in 2025 when coupled with communicating with UAPS or UFOS.

I reached into my back pocket and retrieved my cell phone and started to text Chris with a strange request. "What are you doing?" Amy asked. I looked back disbelieving even myself about my request that I was about to text. I must have resembled a crazy person. "I am going to ask Duncan if he can remote view the backyard right now. Maybe he can see something we can't or confirm my suspicion that a bigfoot just threw a rock at our shed right."

I shook my head in disbelief as I pressed "send". I sent Chris another text with our address to serve as coordinates. Remote viewers were given a longitude and latitude when they were looking into targets. Only to be revealed by removing a photo of the location from a manilla folder to compare the drawing by the remote viewer of the target for accuracy. Did they see the target correctly? I texted Chris with short details of what just happened and that I thought it might be a bigfoot. His response and his quickness to accept this request came within a few seconds. "Okay, hang on". About 45 seconds to a minute later he texted.

Chris texted his response...*"You are on to something..."*

I read the text and looked back at Amy. "No fucking way." I showed her his response. She looked back at me with concern. He then sent another text that I had to reread as I couldn't believe it. I then started looking around with the feeling that whatever was in there...was still there. Chris sends another text.

"*Your suspicions are correct.* You don't have one bigfoot. You have *two*. There is a parental figure standing by a tree watching as a younger sasquatch or juvenile is the one that threw the rock. They are trying to interact with you and your family. And yes, they are *cloaked* - that's why you can't see them."

"Holy shit" I muttered. Is this for *real*?

I started to instantly question if getting into this was a good idea or not. Bigfoot, UFOs and the paranormal. When you look into this stuff, they seem to be aware of your presence or interest and they look back. It also dawned on me that the past few days or so every time we were outside branches or some object was hitting the shed. It was too soon for acorns. I didn't think anything of it or make the connection. It may have been trying to get our attention for a while now. I still wasn't sure if I accepted this answer from Chris Duncan, but I know what he can do - and he's one of the best in the world.

THE RESPONSE

We tend to share all this strangeness with the kids in case they see or experience similar phenomena. They have all seen some things and being closer to the forest was making them a little uneasy. We had a regular pack of coyotes coming through, deer, wild turkeys and black bears were seen close by. There is a horse farm just down the road and sightings of UFOs, orange orbs and bigfoot throughout this area are another commonality.

One of the benefits of having participated in shows such as *Paranormal Caught on Camera* on the Travel Channel and *Expedition Bigfoot* as well as other projects, I get requests to come and speak at various conferences throughout the country and internationally. A few weeks later, I was one of the speakers for the upcoming Michigan Paracon conference and met some new friends, Arjay and Deb DeRousse. They performed a shamanic healing over me while I was at the conference as I discussed with them that I was off and not like myself since returning from Alaska. I also explained what had just happened back at home. After the session, I was finally feeling lighter and clearer and finally had a sense of relief. But I still couldn't help shake what happened to Amy.

I checked in back home. Nothing. No action since I had been gone for two days. I was heading home later that evening and wouldn't arrive until well after midnight. Once I got home, I couldn't sleep. I was wide awake. I couldn't stop thinking about that damn rock being thrown at the shed and Duncan confirming the cloaking sasquatch were to blame. I started to really doubt what was happening. Unable to sleep, I got up and I was in my office doing some writing. I had another "great" idea that came out of nowhere at around 4AM. "Let's see if I can get them to throw a rock at the shed again on my command." Here was the catch. I wasn't going to do this verbally. I was going to send the message *telepathically*.[10] Luckily, no one was up to talk me out of it.

I headed outside onto the deck quietly and entered the darkness. I came a few feet from the edge of the deck where

Amy was originally hanging out on the phone when she had the rock thrown at the shed. It was dead quiet. There was no wind. No sound. But I had the eerie sense something was already aware of my presence and had somehow caught wind of my plan. After a few moments of breathing in the forest air, scanning my surroundings and listening intently for any type of movement, I decided I would try my experiment. Without saying a word, I stared into the pitch-black forest behind our manicured lawn. I could hear the brook running. I walked to the edge of the deck now, my toes hovering over the stairs leading down into my yard.

With all my energy and intention I sent this mental message out into nature in front of me.."Okay, if this is real...I want you to throw a rock at my shed again right now. And I mean right fucking...now...". BOOM! Before I could finish saying "NOW" in my mind...a rock hit the shed so loud it made me jump. Immediately I sensed something that was undeniable. Fear. It was as if someone had poured a bucket of cold water over my head and the fear ran down through every fiber of my being all the way down to my toes.

Seconds later my daughter Lena opened the window of her bedroom on the second floor which faced the backyard and screamed out loud as she saw me on the deck. "Dad - I heard that!" She remembered what we told them about the rock being thrown. "I just asked for that," I replied. I looked back into the forest in disbelief and finally opened my mouth and spoke to whatever or whoever just threw that rock. "Okay - I *get it.* YOU are in charge. I am going to bed."

And with that, I finally went upstairs and went to bed knowing unequivocally without an understanding of how but I just communicated telepathically with something or someone that isn't supposed to exist. They responded with my request within seconds.

What did I get myself into...again?

Chapter Three

THE THREE

"If you want to find the secrets of the universe, think in terms of energy, frequency, and vibration."[11]

Nikola Tesla

I always end my talks with this famous quote by Nikola Tesla. He was beyond his time. Some even believe that he was not of this world. His work continues to fascinate and this quote could indicate some of the answers to the mystery of our reality.

What is energy? **Energy** is a fundamental concept in physics, representing the capacity to do work. Everything in the universe involves energy in some form. Whether it be kinetic, potential, thermal, electromagnetic, etc. When it came to Tesla's research, energy was the key to harnessing and manipulating the forces of nature.

His inventions, such as the Tesla coil and alternating current (AC) systems, aimed to control and transmit energy efficiently across distances. Energy is the element that connects all things. According to the law of conservation of energy, it flows and transforms but it cannot be created nor destroyed. Energy flows through different frequencies, and these frequencies manifest as different vibrations.

Everything is energy.[12] Humans are energetic systems. Modern physics shows that matter is essentially condensed energy. Einstein said that "Mass and energy are different manifestations of the same thing."[13] Atoms are almost empty space.[14] Your body is a dynamic electromagnetic, chemical, and quantum system. Neurons communicate via electrical impulses[15], the heart generates a measurable electromagnetic field, the largest rhythmic electromagnetic field produced by the human body[16] and the brain runs on oscillating electrical rhythms.

Frequency refers to the rate at which a vibration or wave oscillates. In the context of electromagnetic waves, it's the number of cycles per second (measured in Hertz).

Tesla was deeply involved in studying frequencies, especially in the realm of radio waves, wireless transmission, and resonance. He recognized that different frequencies had unique properties and could be harnessed for communication, power transfer, and more.

On a metaphysical level, considering the universe in terms of frequency suggests that everything has a natural vibrational state or frequency. This idea resonates with concepts in quantum mechanics, where particles exhibit wave-like behaviors and are described by probabilistic wave functions.[17]

Tesla believed that the brain works similar to that of a radio and it is a receiver. "The brain is only a receiver, and in the universe there is a core from which we obtain knowledge, strength, and inspiration." The brain doesn't generate consciousness, it accesses the field of consciousness when tuned into the right frequency. Consciousness interacts with all forms of energy. It may operate through dimensions or fields we don't measure as of yet.

So if the body is energy the brain is an energetic interface. Consciousness may be the organizing principle that is embedded and a part of the fabric of the universe. Consciousness is non-local but accessible. The observer becomes an active and instant participant in reality rather than just a witness.[18] Our perception is interaction and what we are observing may be observing us back.

Vibration is the movement or oscillation of particles, waves, or objects. Everything in the physical world vibrates, from atoms in a solid to the light waves we perceive.

Tesla's work often focused on resonant frequencies—the specific vibration rates that would amplify or enhance energy transfer. For example, in his research into wireless power transmission, he explored how matching frequencies could lead to efficient energy transfer without wires.

Philosophically, thinking of vibration as a fundamental aspect of the universe suggests that at the core of all matter and energy are rhythmic, repeating patterns. This idea is mirrored in both classical and quantum physics, where everything from atomic bonds to electromagnetic waves can be understood as vibrational systems.

There are deeper implications when it comes to the unity of energy, frequency and vibration. Tesla's perspective on energy, frequency, and vibration can be viewed as a holistic framework for understanding reality.

The three are interconnected: **Energy** flows through different **frequencies**, and these frequencies manifest as different **vibrations**.

In physics, these concepts are unified in the study of wave phenomena. For example, electromagnetic radiation (light, radio waves, etc.) is characterized by its frequency, which determines its energy and the wavelength (a measure of its vibration). The interplay between energy, frequency, and vibration is evident in fields particularly like quantum mechanics, where particles are described as wavefunctions that have specific energies, frequencies, and vibrational states.

Tesla's ideas align with several principles in modern science:

Quantum Mechanics:

In quantum theory, particles are treated as wavefunctions that vibrate at certain frequencies, and their energy levels are quantized. This connects to Tesla's idea that everything has a frequency and vibration.

String Theory:

In string theory, the fundamental components of the universe are not point particles, but rather vibrating strings.[19] The frequency of a string's vibration determines the particle's properties, among them mass and charge, reflecting the idea that vibration is at the heart of reality.

Resonance in Nature:

The concept of resonance—where systems naturally oscillate at certain frequencies—is fundamental in physics. Resonance is crucial for understanding phenomena ranging from atomic transitions to the operation of electrical circuits, resembling the ones Tesla developed.

Vibrational Healing and Holistic Practices:

During an advanced meditation retreat, Dr. Joe Dispenza, author of the book *Becoming Supernatural: How Common People are Doing the Uncommon*, discusses interdimensional beings making appearances during their meditations.[20] During these multi-day events, Dr. Dispenza started sensing the energy of these beings and feeling their presence as they seemed to join in on these pineal gland focused sessions. These beings seem to be observing these sessions and they were seen lined up along the walls. They are described as being human-like but extremely tall...the size of giants, nine to ten feet tall to be exact! Attendees started to share that they were seeing beings appearing during the sessions and others would corroborate seeing similar beings. Dr. Dispenza was seeing these same beings as well but didn't immediately discuss it.

Dr. Dispenza was getting up for the day during one of these retreats and he started to hear these beings talking to him while he was in a theta-like state between wakefulness and dreaming.[21] They communicated, "we can only lower our frequency to a certain point. We cannot lower our frequency any lower or we would change our consciousness, and we love ourselves too much to do that...but if you can raise the frequency of the audience to a higher frequency, we will meet in that place where that energy intersects." Sounds reminiscent of a doorway or entryway to another place to me, this meeting place that consciousness is the vehicle in which you drive to get there.

Digging further, I am having trouble identifying a description of what these beings look like. Reports from some attendees state that they were seeing "blue beings" that are often associated with healing or considered spiritual guides. Several traditions for instance Kundalini awakening, a spiritual practice rooted in Hinduism, often involve the rise of energy through the chakras, leading to heightened spiritual awareness and experiences that some may interpret as encounters with divine or spirit-like beings.[22]

I recall having a very vivid dream of being led down a hallway by a being, a blue cat-like being that was bipedal and showing or teaching me something. All along the walls of this hallway as we were walking down were lined up on either side with native americans from the past. They acted almost as if they were spiritual guardians or warriors. I was eventually given a tour at a company I would work for prior to leaving to join the Discovery television series Expedition Bigfoot. While we walked down the hallway, I was told that the location was selected because of its spiritual energy.

The lady that was giving me the tour was an empath and a Reiki master and stated that she saw native americans lined up on either side of the walls when they initially visited the building. This was in the city of Littleton, Massachusetts, which is also known as the epicenter of spiritual activity and the location of where the tribes of New England would perform their vision quest.[23] It was considered sacred or special and a few miles from Monsterland. Coincidence?

Why Consciousness Was the Third Rail of Science

In the 20th century, science embraced a worldview that treated consciousness as an accidental byproduct of the brain, simply a chemical hallucination that conveniently allowed us to observe the universe but wasn't important enough to be part of it. I have always believed in mind over matter and that there is a higher power at play in our daily lives.

This chemical view had two advantages:

1. It simplified the world.
2. It avoided hard questions.

But it also created a blind spot. If consciousness is a glitch, then any phenomenon involving perception, awareness, or subjective experience gets written off as unreliable. That's how experiencers — of UFOs, of Sasquatch, of near-death states — were dismissed for decades. Not because their stories were weak but because their stories were consciousness-based.

Science is comfortable with:

- things that can be weighed
- things that can be timed
- things that can be graphed

Science is NOT comfortable with:

- intelligence without bodies
- perception without cause
- communication without words
- beings that appear and vanish

- experiences that change people

But these are the exact characteristics that define the world of high strangeness and Bigfoot, UFOs, and consciousness phenomena all share them. The phenomenon doesn't just appear, it interacts with the observer. More scientists, researchers and explorers are pushing the boundaries now more than ever. It feels as though a wave has crested and it's going to come crashing down in my lifetime. I hear more and more people coming out of their shell and sharing their experiences that they wouldn't have done a year ago.

People often report:

- vivid telepathic impressions
- intuitive warnings
- time dilation
- altered states of awareness
- emotional downloads
- dream-like sequences
- synchronicities leading up to and after encounters

These aren't signs of delusion. They are signatures of something intelligent. Jacques Vallée recognized this in the 1960s and paid dearly for saying it out loud.[24] John Mack identified it in abductees and was nearly stripped of his Harvard credentials.[25] Modern UAP research circles around it constantly but avoids using the word "consciousness."

But experiencers know. People in the woods at 2 a.m. know. The ones who've heard the knocks and the calls in the cold dark know that it is not just physical or biological. It is not just a craft or creature. It is something that engages with the human mind. Most Bigfoot field researchers eventually encounter something that doesn't fit in the biological box. Something that forces them to rethink what they're dealing with:

- The woods go silent instantly.
- Fear hits in the manner of a wave.

- People sense a presence before anything is seen.
- Orbs appear near trackways.
- Structures appear overnight.
- Eye shine with no known ambient light source.
- Heavy footfalls with no tracks.
- Sudden euphoria or calmness.
- Encounters that feel "arranged," meaningful.
- Repeated patterns across unrelated witnesses.

This isn't fantasy. These are data points and consciousness is the only framework that can explain them without denying the experience. Once you admit consciousness can play a role, the Bigfoot phenomenon begins to resemble what Vallée described in UFO encounters: a kind of intelligence that operates in and out of our physical perception, interacting with witnesses selectively, personally, and occasionally symbolically. This is high-strangeness and it is the bridge between Bigfoot and UAP.

Something extraordinary has happened in just a few years, the collapse of the consciousness stigma. Consciousness is no longer considered taboo.

Universities now have:

- consciousness labs
- psi research programs
- psychedelic treatment studies
- medical institutions studying NDEs
- physicists exploring the observer effect
- neuroscientists probing near-death continuity

The Pentagon has studied:

- precognition
- telepathy
- remote viewing[26]
- human–machine mind interaction

NASA has openly discussed UAP as something that "may involve biological intelligence or unknown forms of physics."[27]

And in the Bigfoot world:

- more researchers are speaking about interdimensional possibilities
- experiencers are openly discussing telepathic impressions
- documentaries are no longer afraid to explore the consciousness angle
- orbs, portals, and synchronicities are mentioned without ridicule
- the field is moving beyond the purely biological model

This is not fringe anymore, we have entered into the new frontier. The component that ties Bigfoot, UFOs, and consciousness together is not speculation. It is the patterns and they repeat across culture, centuries, witnesses, and geographies.

When unrelated people in unrelated fields report the same kinds of experiences, something deeper is happening. This brings us to the next stage, the convergence. The Convergence is the moment when Bigfoot, UAP, and consciousness stop being separate subjects and become one phenomenon expressed through different masks. By blending established physics, peer-reviewed research, historical investigation, and first-person experiential data my goal is to identify where the consensus ends and inquiry begins.

The Convergence: Bigfoot, UFOs, and Consciousness as One Phenomenon

There comes a point in every investigator's journey where the walls between Bigfoot, UFOs, and consciousness begin to thin. At first, they seem as though they are entirely separate mysteries—one biological, one technological, one philosophical. But the deeper you go, the more the borders blur. Bigfoot researchers begin noticing strange lights. UFO investigators stumble into cryptid reports. Consciousness researchers encounter phenomena that act the way intelligent observers would. Witnesses who encounter one phenomenon often encounter another.

Patterns repeat.

Locations repeat.

Experiencers repeat.

And in a flash, the once-clear divisions collapse into something far stranger and far more coherent. Bigfoot is not just a creature in the woods. UFOs are not just craft in the sky. Consciousness is not just a brain-generated illusion. They are facets of a larger, interconnected system, a phenomenon that expresses itself through different masks depending on the context, the witness, and the environment.

Researchers call these places window areas, liminal zones, or thin places — locations where the boundary between physical reality and something deeper becomes permeable. These are the places where Bigfoot trackways appear near orb sightings, where hunters see cloaked shapes, where electrical devices fail, where time and emotion distort, where people feel watched before anything announces itself.

These are the places where patterns converge. Across North America and the world, Bigfoot sightings cluster in the same regions as:

- historic UFO flap zones
- military UAP observations
- electromagnetic anomalies
- ancient sacred sites
- missing person clusters
- orb sightings
- portals or vortex claims

These are not random overlaps, they are statistical inevitabilities once you map the data.

Take the following examples:

- Skinwalker Ranch: Bigfoot, UFOs, orbs, portals, and consciousness effects
- Mount Shasta: UFOs, lights, strange humanoids, interdimensional myths
- Kentucky's Pennyroyal: cryptids, orbs, abductions, underground anomalies
- British Columbia: intense Sasquatch activity and frequent UAP reports
- Hudson Valley: classic UFO flap region with multiple Bigfoot encounters

This suggests a common mechanism — not biological, not technological, but environmental and consciousness-linked. Those who see UFOs often see orbs later, encounter paranormal activity in their home. People who see orbs later hear wood knocks or find tracks and ones that encounter Bigfoot often begin experiencing synchronicities.

People who have a consciousness event (NDE, OBE, meditation breakthrough) become more likely to encounter phenomena. This is not because witnesses are "imaginative." It is because they have been opened. A connection has been made and now a signal or signature is being communicated. Once this consciousness barrier breaks, it rarely reseals.

The shamans of old, the initiates or prophets, this isn't new, just a different label. When this connection occurs in the spiritual world it is very likely there is a disconnect in the physical world. Contact or exposure had negative physical effects, revealing encounters can crush careers and ruin relationships as well as shatter belief systems. This convergence suggests that there is an ongoing process to various levels, it doesn't stop, it intensifies.

From Tesla's intuition that the mind is a tuning device, to Dispenza's measurements of altered human states to anomalous intelligences associated with UAP, Bigfoot and the paranormal. What we refer to the unknown and the unexplained may not be distant at all, it may be right here and activated the moment consciousness aligns with it.

Chapter Four

UNVEILING THE INTERDIMENSIONAL ENIGMA

Before examining who they may be, we must examine where reality itself allows for them to exist. There have been recurring intelligences throughout history and patterns of contact. Let's explore the scientific, governmental, mythological, and consciousness-based structures that make interdimensional interaction plausible rather than fantastical.

Physics, mythology, recent intelligence documents related to disclosure and experiencer testimony are additional data streams we can't ignore. What follows is not a claim of certainty, but a convergence of disciplines pointing toward a universe far less isolated, linear, or closed than we were taught.

In February of 2025, House Committee on Oversight and Government Reform Chairman James Comer (R-Ky.) announced the establishment of a task force titled the "Task Force on the Declassification of Federal Secrets." The task force's mission is to examine the declassification of materials in the public interest and will be led by Rep. Anna Paulina Luna (R-Fla.) with UAPS and UFOs falling under this task force.[28]

Congresswoman Anna Paulina Luna shares with Joe Rogan on his podcast in August that "*to break it down into simple terms, some of the tech that exists, whatever these things have, these energy things have...they call them interdimensional beings and I think they can operate in and through the times spaces that we currently have...and that is not something that I came up with on my own, this is based on stuff that we have seen, that based information that we have been told and there is this historical aspect that gets into the deeper theories and concepts of religion and the history that we currently know...what I can tell you without getting into classified conversations that there have been incidences that I believe where very credible people have reported that there has been movement outside of time and space."*[29]

This was a groundbreaking revelation that many didn't seem to notice. In the realm of theoretical physics and metaphysical exploration, the concept of dimensions beyond our perceivable reality has sparked the imagination of many. What if there are hidden dimensions, unseen by the naked eye, coexisting with our own? Could there be a civilization or civilizations existing in a parallel dimension, concealed from the annals of human history? Will we in the very near future be required to reframe our place in a multiverse of unseen realities?

What is reality? To comprehend the notion of hidden dimensions, one must first grasp the fundamentals of our own spatial reality. In our everyday experience, we navigate a three-dimensional world – length, width, and height. However, theories in physics propose the existence of additional dimensions beyond these three.[30]

String theory, a branch of theoretical physics, suggests the existence of at least ten dimensions, with the possibility of even more. These dimensions are compactified, meaning they are intricately coiled or curled up in a way that makes them imperceptible in our everyday encounters.[31] While these dimensions remain elusive to our senses, they could hold the key to understanding the fabric of the universe.

Multiverse Theory and Quantum Revelation

The concept of a multiverse expands upon the idea of hidden dimensions. In the multiverse theory, multiple universes coexist simultaneously, each with its own set of physical laws and constants. This implies that there are parallel realities beyond our own, existing alongside but typically inaccessible to us. What we might interpret in the skies or seas as extraterrestrial could very well be interdimensional in nature. Recent developments are pointing in that direction.

"It lends credence to the notion that quantum computation occurs in many parallel universes, in line with the idea that we live in a multiverse," declared the founder of the Google Quantum AI team, physicist Hartmut Neven. Google physicists reached a milestone with their quantum computing chip called "Willow". Making the announcement on December 17th, 2025, when the chip took only five minutes to solve a computational problem so complex it would have taken today's most advanced super-computers approximately 10 septillion years to crack, which is more than the age of the universe! Timely, this was during the wave of the New Jersey drones as well as throughout the country.

Google's "Willow" is one of their quantum computing chips, designed to advance the capabilities of quantum processors beyond classical computing. Built as a successor to previous quantum chips like "Sycamore," Willow aims to improve qubit coherence and control, allowing for more complex quantum calculations. Quantum computers such as Willow use quantum bits (qubits), instead of 0s and 1s as in classical computing, which can exist in multiple states simultaneously thanks to quantum superposition, enabling exponential processing power compared to traditional bits. Google has been testing these chips in tasks that push the limits of quantum advantage, attempting to demonstrate problems solved faster than any classical computer could.

In the process of experimenting with quantum processors, some Google researchers have speculated based on quantum mechanical behaviors that our reality might be part of a Multiverse. This idea stems from the "Many Worlds Interpretation" of quantum physics, which suggests every quantum decision splits into parallel realities.[32]

When quantum chips perform operations, their outcomes reflect this probabilistic behavior, hinting that countless possibilities may unfold across parallel universes. While this is still philosophical and speculative, Google's advancements in quantum computing have reignited public interest in the idea that we might live in a vast, branching Multiverse.[33]

If we consider the possibility of hidden civilizations, it becomes conceivable that they reside in one of these parallel realities, separated from our own by the veil of different dimensions. Could there be a civilization thriving beyond our sensory perception, living a parallel existence just beyond our grasp? Are they visiting us?

Theoretical physics also entertains the intriguing idea of portals or gateways that connect different dimensions. While this concept might sound akin to science fiction, physicists explore the mathematical viability of wormholes and other cosmic phenomena that could serve as bridges between dimensions. Reality is revealing that it is layered and interactive, not strictly material, not sealed or singular. It has multiple access points when looking at ancient architecture, advanced physics and altered states of consciousness. They all seem to converge with interdimensional implications.

Portals Discovered by NASA

NASA-funded research has uncovered hidden portals in Earth's magnetic field, known as X-points or electron diffusion regions, where Earth's magnetic field connects directly to the Sun's magnetic field. Discovered by Plasma Physicist Jake Scudder from the University of Iowa thirteen years ago in 2012, these portals, located tens of thousands of kilometers from Earth, open and close multiple times daily.[34] It allows energetic particles to flow through and influence our planet's upper atmosphere, leading to phenomena in the manner of geomagnetic storms and auroras.

The discovery of these magnetic portals aligns intriguingly with theories of higher dimensions and the multiverse. In theoretical physics, concepts such as wormholes propose connections between different regions of spacetime or even parallel universes. While Earth's magnetic portals are not wormholes, their existence demonstrates natural, unseen connections facilitated by complex magnetic interactions, suggesting that similar mechanisms could exist on a cosmic scale, potentially linking different dimensions or universes.[35]

This finding supports the plausibility of the multiverse theory, which posits multiple, perhaps infinite, universes existing simultaneously. The natural formation of these magnetic portals indicates that the universe may harbor hidden structures facilitating interactions beyond our current understanding. Exploring these phenomena could provide insights into the fabric of reality and the potential existence of dimensions beyond our observable universe.

Myth and Megastructures: The Architecture of the Hidden

If such things as portals exist, they could provide a means for communication or even travel between our reality and hidden dimensions. Ancient myths and legends from various cultures often depict portals, doorways, or gates to other realms, suggesting that humanity has pondered the existence of such connections for centuries.

It is important to consider archaeological enigmas and anomalies that defy conventional explanations. Throughout history, there have been discoveries that challenge our understanding of the past, leading some to speculate about the existence of advanced civilizations hidden from historical records.

Ancient structures comparable to the Great Pyramid of Giza and Stonehenge, with their precise alignments and mysterious purposes, invite contemplation on the influence of hidden dimensions. Could the builders of these structures have had access to knowledge from a civilization existing in a parallel reality? What if these were generators of some sort to create these portals with a flip of a switch. Just days ago a major announcement made shockwaves across Egyptian archaeology and could be one of the most significant finds in human history.

With the use of Synthetic Aperture Radar (SAR) tomography, scientists Corrado Malanga of the University of Pisa and Filippo Biondi of the University of Strathclyde scanned the Khafre Pyramid and uncovered what appears to be an underground system stretching two kilometers beneath all three major pyramids![36]

The findings, made public through a scientific paper and a press release issued on March 15, 2025 detail five identical multi-level structures connected by geometric passageways near the base of the Khafre Pyramid. There has been a long-held conspiracy that these were energy generators or some sort or a device used for another purpose. Nikola Tesla believed that the pyramids harnessed and converted the earth's natural energy.[37]

The discovery of eight vertical cylindrical wells is accompanied and encircled by spiral pathways descending 648 meters below the surface! These wells ultimately merge into two massive cube-shaped chambers with each measuring 80 meters on each side." These structures are twice the height of the tallest building on earth! It could be one of the most important discoveries in human history and may help define who we really are. What is or was the pyramid's true purpose? Rumors have circulated about reports of a concealed underground pyramid in Alaska, often referred to as the "Black Pyramid", alleged to possess significant energy capabilities and to be under government secrecy. A location also known as the Alaskan Triangle where people have gone missing which ties in with portal locations.

Theories of its capabilities range from emitting powerful electromagnetic fields, potentially influencing or manipulating human consciousness is another. Linda Moulton Howe, an investigative journalist, has reported on accounts from military personnel regarding an alleged underground pyramid in Alaska. For instance, retired U.S. Army counterintelligence warrant officer Douglas Alan Mutschler claimed that a massive subterranean pyramid, larger than Egypt's Great Pyramid, was discovered in Alaska and possesses significant energy capabilities. He stated that the structure "had the ability to power all of Alaska and Canada" and was considered "more secret than the Manhattan Project'![38]

Mythology and Folklore: Echoes of Hidden Realities

Myths and folklore from various cultures often feature tales of gods, beings, or entire civilizations residing in realms beyond the reach of mortal eyes. These stories, once dismissed as imaginative narratives, are being discussed in more depth. One strong theory is that the UFO/UAP enigma could be emerging from, entering into and residing with our oceans. Stories of Atlantis flood to mind. Could they be hiding in our oceans or could these areas be where some of these portals are secretly and conveniently located?

The ancient Greek concept of Mount Olympus as the dwelling place of the gods or the Norse realms of Asgard and Midgard are examples of mythological landscapes that hint at the existence of hidden civilizations. Could these stories be echoes of encounters with beings from other dimensions?

Some researchers suggest that ancient maps contain clues about hidden dimensions and unknown civilizations. Piri Reis Map, an early 16th-century Ottoman map, is often cited as an example. The map displays an accurate representation of the coastline of South America and Antarctica, areas not officially discovered until centuries later. How did the cartographers of the time possess such precise knowledge? Could they have accessed information from a hidden civilization through some form of dimensional communication or advanced technology unbeknownst to us?

The CIA's 1983 report titled "Analysis and Assessment of Gateway Process" explores the Gateway Experience, a program developed by the Monroe Institute aimed at expanding human consciousness.[39] Central to this process is the use of "Hemi-Sync," a technique that synchronizes the brain's hemispheres through specific audio stimuli, facilitating altered states of awareness. The report suggests that by achieving such synchronization, individuals can transcend the limitations of time and space, accessing a universal holographic energy field.

The document posits that human consciousness, when properly attuned, can resonate with the Earth's electromagnetic field, creating a powerful carrier wave. This resonance purportedly enables the mind to access and interact with higher dimensions or other planes of reality. Such interactions imply that consciousness is not confined to the physical body but can extend into interdimensional realms, aligning with theories that view the mind as inherently interdimensional.

Furthermore, the report discusses the concept of the universe as a holographic projection, where each part contains the whole. In this framework, human consciousness can tap into this universal hologram, suggesting that the mind has the potential to access information beyond conventional sensory experiences. This perspective supports the notion that our minds possess interdimensional capabilities, capable of transcending physical reality to connect with broader, non-material dimensions.[40]

Many spiritual traditions and esoteric teachings propose the existence of hidden dimensions and realms accessible through spiritual practices. Practices such as meditation, astral projection, and shamanic journeying are believed to open gateways to other dimensions.

In these traditions, the hidden civilization might be considered as beings from higher planes of existence, guiding humanity from behind the veil. Could the knowledge bestowed upon ancient mystics and seers have originated from a hidden civilization communicating across dimensions?

Across physics, archaeology, intelligence research, and ancient myth, the same implication emerges: reality appears layered, permeable, and participatory. Whether accessed through technology, consciousness, or forgotten knowledge, these interfaces suggest humanity has brushed against other realms before—and may be doing so again now.

Chapter Five

THE ORIGIN OF CONSCIOUSNESS

The nature of human consciousness has long intrigued scientists and philosophers, leading to various theories about its origins and mechanisms. One compelling perspective suggests that consciousness may stem from dimensions beyond our conventional understanding. This notion states that our conscious experience is not merely a byproduct of neural activity but is influenced by, or even originates from, higher-dimensional realms. Modern physics increasingly suggests that reality is not built from solid matter, but from information structured through fields, energy, and light.

At the smallest measurable scales, particles behave less like objects and more like probability distributions until they are observed or interacted with.

A notable example of this idea is the Orchestrated Objective Reduction (Orch OR) theory, developed by physicist Roger Penrose and anesthesiologist Stuart Hameroff. This theory proposes that consciousness arises from quantum processes within the brain's microtubules, suggesting a connection between brain activity and fundamental aspects of the universe. According to Orch OR, these quantum processes are influenced by spacetime geometry, implying that consciousness is linked to the fabric of reality itself.[41]

Similarly, the holonomic brain theory, introduced by neuroscientist Karl Pribram in collaboration with physicist David Bohm, likens brain function to a holographic storage network.[42] This model suggests that information in the brain is distributed in a manner akin to holographic patterns, allowing each part of the brain to contain the whole of stored information. Such a framework implies that consciousness emerges from complex, non-local interactions within the brain, resonating with the concept of a holographic universe where each part reflects the whole as is mentioned in the now declassified CIA document.

Recent neuroscience discoveries reveal that the human brain contains microscopic crystals of magnetite and calcite. These materials are known for their sensitivity to electromagnetic fields.[43][44] These biogenic crystals, found particularly in regions associated with perception and orientation, suggest the brain may be naturally equipped to detect or respond to subtle energetic fluctuations beyond conventional sensory input. Some researchers have proposed that such structures could act as biological transducers, converting electromagnetic or quantum-level information into neural signals.[45] Within an interdimensional framework, this raises the possibility that human consciousness is not merely a byproduct of the brain, but an interface capable of interacting with deeper layers of reality.

These theories collectively challenge the traditional view that consciousness is solely a product of classical neural interactions. By introducing quantum mechanics and higher-dimensional considerations, they open the door to the possibility that our conscious experience is deeply intertwined with, and perhaps even originates from, dimensions beyond our immediate perception.

The reality as we know it is that the human eye can only see a small percentage of the entire electromagnetic spectrum, 0.0035%.[46] So infrared, UV, X-rays, gamma rays and radio waves are invisible to us. In comparison the ordinary matter (protons, electrons and neutrons) we can see and touch only makes up 5% of the known universe. Dark matter (27%) and dark energy is invisible and largely a mystery.[47]

We infer the existence of them both through their effects as gravitational lensing, galaxy rotation curves and cosmic expansion rates but we cannot observe dark matter and energy directly. In a similar way, this is how we observe consciousness through behavior and brain activity without really understanding its underlying mechanism and its nature. Is it in our head, is it around us as an aura or is consciousness in everything?

Dr. Michael Pravica, a physics professor at the University of Nevada, Las Vegas, has proposed a theory in 2024 that suggests human consciousness may originate from dimensions beyond the four we commonly perceive—length, width, height, and time. This concept, rooted in hyperdimensionality, posits that the universe encompasses additional dimensions that are typically beyond our sensory perception.

Pravica suggests that during moments of heightened awareness—such as engaging in creative activities, scientific exploration, or dreaming—our consciousness might transcend these physical limitations and access these hidden dimensions, leading to profound insights and inspiration. What instantly comes to mind is what the shamans discuss when mentioning the dream world is the "real" world in which we leave our body while in the dream state visiting various worlds and dimensions returning when we awaken from our slumber.

To illustrate this idea, Pravica uses the analogy of a two-dimensional being living in a flat world. Such a being would be incapable of perceiving three-dimensional objects in their entirety, only experiencing cross-sections as they intersect its plane of existence. Similarly, humans operating within a four-dimensional framework may be unaware of higher dimensions that exist around us. Pravica argues that just as the 2D being cannot fully comprehend the 3D world, we might be limited in perceiving the full spectrum of dimensions in our universe.

The Mandela Effect

The Mandela Effect, a phenomenon where a group of people recalls an event or detail differently than the official historical record, raises questions about the nature of reality and the possibility of dimensional shifts.[48] While skeptics attribute this to false memories, proponents of the Mandela Effect propose that it could be evidence of individuals shifting between parallel realities. If the Mandela Effect is indeed a result of dimensional shifts, could it be that hidden civilizations play a role in these fluctuations, influencing our collective memories from behind the veil?

I have memories of seeing the television news report of Nelson Mandela dying while being held in prison, which was not the case. He later became President. The "Berenstein" Bears books, "Jiffy" Peanut Butter and Fruit of the Loom logo with a cornucopia among others, I have "memory" of or a sense of knowing that it once was...real. The Mandela Effect reveal is considered to be slight differences but memories of another reality with suppressed recall. They are slight alterations to reality or rather different versions of probability that then get shifted.

Could this Mandela Effect also be an indication that we have learned how to manipulate time so that our reality can shift. What once was can be changed? Or do the "others" know how to do this and are continually doing it now? I have a very weird feeling that we have accomplished time travel and there are also theories that the aliens themselves, the ships, are not spaceships but time machines.

A recent study focusing on the Mandela Effect by the University of Chicago has provided scientific evidence supporting the existence of this phenomenon, confirming that these vivid yet "inaccurate memories" are a shared human experience. In the study, researchers presented participants with 40 well-known logos and characters, each accompanied by two altered versions with one reflecting common misrememberings and another with unrelated changes.

Participants frequently selected the incorrect images with high confidence. For instance, many believed Pikachu has a black-tipped tail and that the Monopoly mascot, Rich Uncle Pennybags, wears a monocle, though neither feature exists. But I swear I can *see* that monocle from my youth playing that game.

Assistant Professor Wilma Bainbridge, a co-author of the study, highlighted the significance of these findings, noting that consistent patterns of false memories across populations demonstrate the fascinating nature of the Mandela Effect. But are these "false" memories witnessed as time shifts for people that are more in line with the frequency or are more aware of these subtle differences that hold a strange pit in their stomach, a sense of metacognition or sharing of the truth?

The study also explored potential explanations for the Mandela Effect, such as schema theory which is the idea that people fill in memory gaps with familiar associations. However, this theory couldn't account for certain cases, like the frequent addition of a cornucopia to the Fruit of the Loom logo, suggesting more complex cognitive processes are involved.

These findings from the study have broad implications, raising questions about the reliability of memory in contexts, particularly eyewitness testimony and the spread of misinformation. But why would a massive amount of people remember something false? Doesn't it remind you of the UFO encounter and abduction scenarios where reality is blurred, where we are meant to forget the experience?

In my opinion, if the multiverse is real, then dimensions are real and that immediately means that theories such as the Mandela Effect have significance and could very well be indicators of these dimensions as alternate realities or realms.

Could realities or timelines be shifted? I think they already have.

Here is what what Remote Viewer #001 for the CIA Joseph McConeagle says about "reality" during the Shawn Ryan Show.[49] Theoretical physicist Michio Kaku echoes the same explanation.[50] And Einstein describes our universe as living on the edge of a bubble that is expanding away from other universes.[51] McConeagle explains what he thinks reality really is.

"Your world is where you are and what you are doing. And that's REAL...only while you are doing it in that place at that time. I look at reality as a huge 10 x 10 room full of bubbles. So every decision I make I am moving from one bubble to another one and it might be a bubble that is not a tangent bubble but another one that is somewhere in the room of bubbles. So reality for me is a massive bubble of possibilities where we all go different ways for different reasons and we collect experiences that are somewhat similar maybe but they are never the same."

So in essence with each decision or thought we make we are selecting an endless amount of bubbles or realities. These realities or dimensions could be right next to us or above us. Every thought, decision or action you take will shift your reality to another reality.

The exploration of hidden dimensions and civilizations requires a collaborative effort that bridges the gap between scientific inquiry and mystical traditions. By combining the methodologies of physics, metaphysics, archaeology, and consciousness studies, researchers may uncover patterns and correlations that elude isolated disciplines. This interdisciplinary approach acknowledges the complexity of the subject and seeks a comprehensive understanding that integrates diverse perspectives. More collaborations and research are being harnessed in these areas of study. Some have been classified.

The exploration of hidden civilizations is often intertwined with discussions about extraterrestrial life. Some theories propose that extraterrestrial beings could originate from dimensions beyond our own, making them interdimensional rather than strictly extraterrestrial.

If extraterrestrial beings have the ability to traverse dimensions, they might be part of the hidden civilization that interacts with our reality intermittently. Could UFO sightings and encounters with extraterrestrials be glimpses into the activities of beings from parallel dimensions? In addition to spatial dimensions, time is considered by physicists as a dimension. The concept of time as a dimension implies that past, present, and future exist simultaneously. If this is the case, could there be a hidden civilization that transcends temporal constraints, residing in a perpetual state where time converges?

Ancient prophecies and mystical teachings often allude to the cyclical nature of time. Could these be hints of a civilization that understands time as a fluid dimension rather than a linear progression? Quantum mechanics, the branch of physics that studies the behavior of particles at the quantum level, introduces phenomena including entanglement and superposition. These phenomena suggest that particles can exist in multiple states simultaneously and be connected regardless of distance. What Einstein called “Spooky action at a distance”. Could the principles of quantum mechanics extend to larger scales, allowing for the existence of hidden dimensions and civilizations that operate beyond the constraints of classical physics?

Psychonauts, individuals who explore altered states of consciousness through practices such as psychedelics or meditation, often report encounters with entities or realms that transcend ordinary perception. These experiences, while subjective, raise questions about the nature of consciousness and its potential connection to hidden dimensions.

Could psychonautic journeys provide a glimpse into the hidden civilization, accessed through altered states of consciousness? Apparently Remote Viewing is one of these methods to access these realms or realities at will.Entities like Space Force have been training soldiers on how to use this skill, but it has been going on longer than that. We are learning more and more that those that are connected to these secret UFO and UAP crash retrieval programs most if not all have this capability of remote viewing.

Is Remote Viewing the technique that our pilots use in order to speak to the occupants of the UFOs by using their minds as the communication method? Could they even be luring in these strange lights with this capability?

Chapter Six

DREAMS, REMOTE VIEWING AND CONSCIOUSNESS

Illuminating the Mind: Biophotons and the Mystery of Consciousness

If reality is layered with unseen dimensions, then perhaps the greatest instrument for exploring them isn't a spaceship, it could very well be the human mind. In recent years, scientific exploration into the mysteries of the human brain has unveiled a captivating revelation with the existence of biophotons, subtle particles of light emitted by living cells. These minuscule light particles, once considered mere byproducts of cellular processes, are now captivating the attention of neuroscientists and consciousness researchers alike. The profound implications of the discovery suggest a potential link between biophotons and the enigmatic realm of human consciousness.

Historically, the brain was thought to be an organ solely driven by electrical impulses and chemical reactions. However, recent advancements in technology, such as highly sensitive photon detectors, have enabled scientists to detect and measure the presence of biophotons in the brain. Researchers have found that not only does the brain emit biophotons, but their intensity varies in correlation with neural activities. This discovery challenges traditional views of brain function and opens a new frontier in understanding the intricate relationship between light emissions and cognitive processes.[52]

The exploration of biophotons has led scientists to delve into the tantalizing question of whether these light particles play a role in the phenomenon of consciousness. Some theories propose that biophotons may serve as carriers of information within the brain, participating in the complex web of communication between neurons. As consciousness is intricately tied to information processing and integration, the presence of biophotons sparks speculation about their potential involvement in the emergence of self-awareness and subjective experience.

Light functions as a physical phenomena, a carrier of information and a boundary condition of light itself. Photons have no rest mass, experience no passage of time, and exist only in motion, making light less a "thing" and more a process by which the universe communicates with itself. [53]

Just last week, March 10, 2025, a recent breakthrough in quantum physics as Italian scientists have successfully demonstrated that light can behave as a "supersolid," a rare state of matter exhibiting both solid-like structure and frictionless flow.If light can be frozen and manipulated into a solid state, it opens up fascinating questions about how light, energy, and information interact core elements in many interdimensional and consciousness-related theories.[54]

Most researchers believe that UFOs do not follow traditional propulsion systems, using either what is called zero point energy, utilizing light, magnetism, or even consciousness as navigational tools through dimensions. Solidified light could hint at states of matter or technology used by non-human intelligences. A recent theory that has emerged from "whistleblowers" is that the craft do not have lights but what eyewitnesses are seeing is the warping of space time from these vehicles in the form of a light, an after effect not unlike the shadows we create.

In consciousness studies, light is often metaphorically and sometimes literally tied to awareness and perception. If we can control light in new ways, it might one day allow direct interfaces with consciousness, perhaps even enabling technologies like telepathic communication or dimensional exploration tied to altered states. It blurs the line between material and immaterial much like many experiencers describe UAPs and beings interacting with reality in nonlinear ways. This is the new frontier and an old one coming around full circle, Spirit Science.

Quantum Consciousness Hypothesis:

A subset of researchers is exploring the connection between biophotons and quantum physics in the quest to unravel the mysteries of consciousness. The Quantum Consciousness Hypothesis suggests that quantum processes, including the emission and absorption of biophotons, may underlie the nature of subjective experience. Proponents of this hypothesis propose that the brain's microtubules, structures within neurons, may function as quantum processors capable of harnessing the information carried by biophotons.

In 2017, The Blue Brain Project has made a groundbreaking discovery that reveals our brain functions in far more complex ways than previously imagined. By applying advanced mathematics, specifically algebraic topology, the team uncovered multi-dimensional structures in the brain’s networks, with some formations extending up to eleven dimensions![55]

These structures are formed when neurons connect in tightly integrated groups known as cliques, creating geometric patterns that go beyond our standard understanding of three-dimensional space. Sounds similar to sacred geometry.

This discovery not only deepens our understanding of the brain’s architecture but also opens up a radical perspective on the nature of consciousness and perception. The presence of these multi-dimensional structures suggests the brain may process information through hidden dimensions, making it difficult to fully grasp using traditional mathematical or scientific approaches. Henry Markram, director of the Blue Brain Project, noted that the failure of standard network mathematics to detect these patterns may be one reason the brain has remained so elusive in neuroscience.[56]

These findings also have profound implications for our understanding of reality itself. If the brain is operating in higher dimensions, it may indicate that human consciousness is capable of interacting with or perceiving aspects of reality that exist beyond our familiar three-dimensional experience.

This aligns with theoretical physics concepts such as string theory and the multiverse, and suggests the possibility that what we experience as intuition, dreams, or even encounters with UAPs could be glimpses into other dimensions that our multidimensional brains are somehow able to tune into. Maybe these are dimensions from which we come from, go to and will return to when we "die"?

Dimensional Dreaming

The concept of experiencing alternate realities through dreams has intrigued both scientists and philosophers, leading to various theories and hypotheses. One such theory is the "Mag Theory," which proposes that during unconscious states, for instance dreaming, the human brain experiences an altered perception of time. This alteration potentially facilitates connections with other dimensions or alternate versions of oneself, aiming to explain phenomena such as precognitive dreams and unrecognized dream events. The theory suggests that sensory isolation during dreams allows access to information beyond conventional time and reality constraints.

Similarly, the "Simulation Hypothesis" posits that our reality could be an artificial simulation, such as a computer-generated environment. This hypothesis raises questions about the nature of consciousness and whether our experiences, including dreams, are part of a simulated construct. The hypothesis draws parallels between the immersive nature of dreams and the possibility that our waking reality might also be a form of simulation. Where is this reality being projected from? Another reality or our own? Maybe these pyramids found throughout the planet literally had a higher purpose?

Philosopher Antti Revonsuo introduced the "Threat Simulation Theory," which suggests that dreaming serves an evolutionary function by simulating threatening events. This allows individuals to rehearse fight-or-flight responses, thereby enhancing survival skills. According to this theory, the realistic and immersive nature of dreams enables the brain to create scenarios that prepare individuals for potential dangers in waking life.[57]

These theories collectively explore the profound connection between dreams and alternate realities, suggesting that our dream experiences might extend beyond mere reflections of our waking life. They propose that dreams could serve as gateways to different dimensions, simulations, or evolutionary tools, offering deeper insights into the nature of consciousness and reality.

On October 13th, 2024, in a groundbreaking experiment by California-based startup REMspace, where two individuals successfully communicated with one another while in separate lucid dreams—an unprecedented achievement in dream science. This marks the first known instance of bidirectional communication between lucid dreamers. REMspace's founder, Michael Raduga, described the development as revolutionary, likening it to a shift from science fiction into everyday reality.[57]

The experiment used brain-monitoring equipment, a Wi-Fi server, and audio transmission to detect when the participants entered the lucid REM sleep phase. Once the first participant achieved lucidity, a server-generated word in a unique language was transmitted to them via earbuds. The dreamer repeated the word within the dream, and the data was stored. When the second participant entered a lucid state eight minutes later, the same word was transmitted and repeated aloud after waking—confirming that communication occurred during their dreams.

Lucid dreaming, a state in which the dreamer becomes aware they are dreaming, was key to the experiment. It allowed both participants to consciously engage with the communication process within their dreams. REMspace capitalized on this state by precisely timing the message transfer based on brainwave patterns. Though details about the full technology remain undisclosed, the proof-of-concept demonstrates that controlled interaction in dreams is now scientifically viable.

The potential implications are vast. If replicated, this breakthrough could lead to applications in mental health therapy, such as treating PTSD by enabling people to consciously confront trauma in a dream environment. Additionally, it could provide a platform for skill development, problem-solving, and even creative collaboration in ways not possible while awake. Raduga believes such dream-based communication will eventually become as normal as texting or emailing.

Raduga, known for his unconventional explorations of consciousness—including an attempt to control dreams with a brain-implanted chip, continues to push the boundaries of dream science. Although more research is required for peer-reviewed validation, this successful experiment could redefine the relationship between sleep, consciousness, and communication. If proven consistently effective, it may open up entire industries centered on shared dream environments and conscious exploration of the subconscious mind.

Remote Viewing

Remote Viewing is a human perceptual ability to access, by mental means alone, information blocked from normal perception by distance, shielding, or time. You will hear the term psionics being used as the militarized label of this practice. It is an ability that I believe we all have the capacity for, just like painting, drawing or singing for example. Everyone can do it, but you only have a few Picassos, DaVinci or John Lennons out there in the world that can do it really well. When it comes to remote viewing, Chris Duncan is one of those guys that is above average.

The U.S. government's decades-long involvement in remote viewing—essentially psychic spying began during the Cold War where the U.S. and Soviet Union both invested in extrasensory perception (ESP) for intelligence gathering.[58]

One of the strangest and most compelling elements of the program was its foundation in yogic and ancient spiritual traditions. There are parallels between remote viewing and the psychic powers described in Patanjali's Yoga Sutras from ancient India, noting that U.S. remote viewing techniques mirrored those ancient teachings.

This link between modern military operations and esoteric spiritual practices adds a surreal dimension, implying that hidden capacities of the mind long known to mystics are being weaponized by governments, including our own. The parallels between the U.S. government's remote viewing programs and ancient spiritual traditions—particularly Patanjali's Yoga Sutras—are striking and layered:

1. Siddhis (Psychic Powers): In the Yoga Sutras, Patanjali describes siddhis, or extraordinary abilities gained through deep meditation and mental discipline. These include clairvoyance, telepathy, and perceiving distant or hidden objects—all of which mirror the goals of remote viewing. The U.S. remote viewers trained to mentally "see" targets far away bear resemblance to yogic practitioners achieving similar feats through focused awareness.

2. Mental Stillness as a Gateway: Both traditions emphasize quieting the mind to access hidden knowledge. In yogic practice, this is done through samadhi (deep meditative absorption), while in remote viewing, viewers enter a calm, trance-like state to allow impressions to emerge without interference. The belief in both systems is that perception is not limited to the physical senses but is instead expanded when mental noise is reduced.

3. Non-local Consciousness: Yogic philosophy suggests that consciousness transcends space and time—a belief echoed in remote viewing results where individuals perceive locations or events with no sensory contact. This concept of a non-local mind accessing information beyond linear time or physical distance supports the idea that both systems draw from the same metaphysical understanding of reality.

Remote Viewers, sometimes located in sealed rooms, were able to describe distant military installations, terrain, and even classified activities with remarkable accuracy—sometimes up to 100%. In one case, psychic Pat Price described a swimming pool complex from miles away with near-perfect detail. Another viewer, Joe McMoneagle, remote view #001, was able to recount the movements of a target at the Lawrence Livermore National Lab with high statistical reliability. Even stranger, Price once described objects at a site that didn't exist—until decades later it was revealed that it had existed in the past, suggesting not just clairvoyance, but possible retrocognition or access to non-linear time! Time traveling with your mind!

Luis "Lue" Elizondo, a former U.S. Army intelligence officer, gained prominence for his role as the director of the Advanced Aerospace Threat Identification Program (AATIP), a Pentagon initiative that investigated Unidentified Aerial Phenomena (UAPs). He is one of the many "whistleblowers" that are emerging during this age of UFO disclosure.

In August 2024, he published his memoir, ***Imminent: Inside the Pentagon's Hunt for UFOs,*** providing an insider's perspective on the U.S. government's efforts to understand these mysterious occurrences. In his book, Elizondo discusses his experiences and the profound implications of UAPs on national security.

Elizondo has also addressed the topic of remote viewing. In his memoir, he recounts participating in a military remote viewing program under the guidance of Dr. Harold "Hal" Puthoff, a physicist known for his work in this field. Elizondo describes an instance where he and four other agents purportedly used remote viewing techniques to influence a terrorist's dreams, leading the individual to report the experience through his lawyer. He would later reveal that he and David Grusch worked together at Space Force! And guess what, David can do it too, he can remote view.

So Remote Viewing and UAPs have a connecting thread? On July 26, 2023, David Grusch's testimony before Congress marked a pivotal moment in the U.S. government's handling of Unidentified Anomalous Phenomena (UAP).

As a former intelligence officer with the National Geospatial-Intelligence Agency, Grusch claimed under oath that the U.S. has engaged in a decades-long, top-secret program to recover and reverse-engineer non-human craft. He further stated that this program operated without proper congressional oversight and that he had faced retaliation for coming forward. His testimony drew attention not only because of its content, but also because he filed an official whistleblower complaint through the Intelligence Community Inspector General, giving his claims legal weight.

Grusch asserted that the U.S. possesses "intact and partially intact" craft of non-human origin and that "biologics" which he described as non-human were recovered from crash sites. Though he could not share specific evidence in public due to classification laws, he emphasized that he had provided detailed documentation and names of witnesses to the Inspector General and congressional intelligence committees.

His allegations suggest that elements within the government have concealed the existence of UAP recovery programs from both elected officials and the public, raising serious questions about transparency and national security.

The hearing also featured testimony from former Navy pilots Ryan Graves and David Fravor, who supported Grusch's claims by detailing their own encounters with highly advanced aerial objects. Their statements reinforced concerns that UAPs exhibit flight capabilities far beyond current human technology and could pose threats to aviation and military operations. Remote viewing gives our military or operators "insight" to what they can't control, predict or maneuver around.

Another significant whistleblower to recently come forward in 2025 is Jake Barber, a former U.S. Air Force veteran. During a NewsNation interview he revealed his involvement in a covert UFO crash retrieval program. Serving as a helicopter pilot contracted by the military, Barber recounted missions where he and his team were dispatched to recover downed aircraft, some of which he described as "extraordinary and anomalous." In one notable instance, Barber reported observing and ultimately retrieving a white, egg-shaped object lacking any visible engine. This object was one in which he believed to be of non-human origin. Protocols were apparently changed leading up to the recovery of the object. He interpreted this change as an indicator that they were recovering something different and they didn't want to take any chances with adverse effects that this object may have had on the special forces team below that would help in the recovery effort.

Barber's testimony is bolstered by video evidence and corroborated by three other veterans, lending credibility to his claims. He noted that high-ranking officials later confirmed the non-human nature of the technology they encountered. These revelations align with previous disclosures by other whistleblowers, such as David Grusch, and have contributed to ongoing discussions and congressional hearings regarding the existence of secret government UFO programs.

In addition to his crash retrieval duties, Barber highlighted the use of remote viewing, or psionics, in their operations. This technique which we now know has been and is still being used by our military involves perceiving distant or unseen targets through extrasensory means, suggesting that such unconventional methods were integrated into their normal mission protocols.

Barber's account underscores the potential intersection of advanced human cognitive abilities with unidentified aerial phenomena, prompting further exploration into the methodologies employed within these clandestine programs. He recently spoke about meditation during a News Nation interview and how they can call in these craft with their minds.

He has a hard time using the term "summoning" but that is essentially what is occurring. "You can say what you want about Dr. Greer but the guy has been 30 years ahead of anyone else on this subject. The one thing he gets the most shit for, is the one thing he's probably most right about and that's his CE-5 protocol." Using the tool of group meditation, Greer uses this technique to bring in their various craft during their group meditations and attempted communication with occupants of these UFOs or UAPS.

Mind-Reach: Scientists Look at Psychic Abilities by Russell Targ and Harold E. Puthoff is a pioneering work published in 1977 that explores the scientific investigation of psychic phenomena, particularly remote viewing.[59] Yes, the same Puthoff, the legend himself that worked with Elizondo. The book documents a series of experiments conducted at the Stanford Research Institute (SRI) during the 1970s, where the authors worked under government contracts to test whether individuals could accurately describe distant or hidden locations using only their minds. These experiments involved both trained and untrained participants, including military personnel and natural psychics, with many showing statistically significant results.

Targ and Puthoff argued decades ago that psychic abilities such as remote viewing are real, reproducible phenomena that can be studied using rigorous scientific methods. Their research was structured around double-blind protocols and detailed reporting to rule out chance, fraud, or sensory cues. The book highlights several compelling examples where viewers described locations, objects, or events they could not have known through normal means—sometimes in astonishing detail. Their efforts laid the groundwork for further military and intelligence interest in psychic functioning, eventually leading to programs including Stargate and others.

The biggest takeaways from Mind-Reach are that remote viewing is not limited to a few gifted individuals; rather, it appears to be a *latent human ability* that can be enhanced through practice. Just like working out a muscle and having a routine, you can strengthen that muscle. The authors propose that consciousness may not be constrained by space and time, challenging traditional materialist views of the mind.

Their findings suggest the mind can access non-local information, which has profound implications not only for science and intelligence but also for how we understand reality, dimensions and consciousness.

Chapter Seven

A UFO MIND

A New York Post article back in 2021 with Dr. Gary Nolan entitled *A Look Inside People's Brains Who Say They've Had A UFO Encounter* discusses a study focused on individuals who claim to have had encounters with UFOs or UAPs, and the neurological anomalies observed in their brains afterwards.[60] Researchers found that many of these individuals showed unusual brain activity or structural differences, particularly in the caudate-putamen region—an area associated with habit formation, learning, and processing complex information.

Dr. Garry Nolan, a Stanford professor of pathology, led the research and noted that many of these individuals were high-functioning and often held roles in the military, aerospace, or intelligence fields. He described their brain scans as showing enhanced connectivity, similar to that seen in gifted individuals, rather than any signs of mental illness or damage.[61]

The implications are fascinating: rather than being delusional, some experiencers might have unique cognitive or neurological traits that allow them to perceive or interact with phenomena the general population cannot. This research adds a new dimension to the UFO conversation and that there is a strong chance that it is biological in origin as well.

Uri Geller, an Israeli-British citizen also participated in Project Stargate is renowned for his spoon-bending feats, remote viewing, telepathic skills and claims of unbound psychic abilities. He recently underwent a brain scan in October of 2024 that revealed distinctive features in his brain structure.[62] These anomalies, particularly in regions associated with sensory perception and motor control, may offer insights into the neurological basis of his purported abilities.

Geller has long attributed his psychic talents to an encounter with a mysterious sphere of light during his childhood in Israel. He believes this experience was of extraterrestrial origin, suggesting that his abilities might be linked to alien intervention of some sort.

The brain scan findings have sparked discussions among scientists and skeptics alike. While some researchers consider the possibility that Geller's unique brain structure could contribute to heightened sensory or cognitive functions, more scientists are looking into what the ancients would call "Spirit Science".

Telepathy

In 2024 a new podcast emerged that shot straight up to number one on the charts. Dr. Diane Hennacy Powell, a neuroscientist and psychiatrist, has been exploring the potential for telepathic communication, particularly among nonverbal autistic individuals. Her work gained significant attention through *The Telepathy Tapes*, a podcast series that documents her research and interactions with families worldwide. The series presents compelling accounts and experiments suggesting that some nonverbal autistic individuals can accurately convey information without traditional forms of communication.[63]

In these experiments, participants, often assisted by facilitators, are asked to identify numbers, words, or images known only to the facilitator or experimenter.[64] The accuracy and specificity of their responses have led some to consider the possibility of telepathic abilities.

Losing her medical license in pursuit of this groundbreaking research, she is working to prove her sanity and reclaim her medical profession. This is how paradigm shifts happen. Dr. Powell's research challenges conventional scientific understanding, proposing that these individuals might access information beyond the known sensory channels. But where is it coming from? If telepathy exists naturally in some humans, BrainNet suggests it may also be artificially replicated—implying the brain is already compatible with direct signal exchange.

Announced in February 2025, a groundbreaking study by scientists at the University of Washington, who have made major progress toward enabling telepathic communication between humans. Using a combination of electroencephalography (EEG) to record brain signals and transcranial magnetic stimulation (TMS) to deliver signals, the team developed a system they call "BrainNet."[65] This system allows individuals to exchange simple thoughts, such as "yes" or "no" responses, from one brain directly to another without the use of traditional speech, typing, or gestures.

In one experiment, three participants were connected through this brain-to-brain interface to play a collaborative Tetris-like game. One participant—the "receiver"—had to rely entirely on brain signals sent from two "senders" to decide whether to rotate a block. The process worked by decoding the sender's brain signals through EEG and then delivering the message to the receiver's brain using TMS, which stimulates specific brain areas associated with visual perception. The results showed the receivers could make accurate decisions based solely on the transmitted brain signals, suggesting a successful exchange of intent or thought.[66]

While the communication is still limited to basic binary decisions and operates through external hardware, it marks a significant step toward direct neural interaction. Researchers believe these brain-to-brain interfaces could eventually enable more complex exchanges, potentially allowing for silent communication between people or group problem-solving without verbal discussion. The technology may also offer a new way for individuals with speech or movement disorders to communicate effectively.

The broader implications of this work reach into the philosophical and metaphysical realms. If telepathy—defined as mind-to-mind communication—is possible even in a technological form, it challenges long-standing beliefs about the separation between individuals and the limitations of consciousness. The research lends scientific credibility to concepts once considered fringe, such as psychic communication or non-verbal intuition. It also suggests that consciousness might be more fluid and interconnected than previously assumed.

Ultimately, this breakthrough opens new frontiers in neuroscience, human communication, and unequivocally the origin of consciousness itself. While still in its infancy, the BrainNet system offers a glimpse into a future where human minds could connect directly, potentially transforming how we collaborate, teach, empathize, and understand one another or the others. The study not only showcases the power of technology but also hints at hidden dimensions of the human experience that science is just beginning to uncover.

In the updated Moment of Contact: New Evidence Revealed documentary filmmaker James Fox, who I know on a personal level, interviewed a Brazilian neurosurgeon (Dr. Italo Venturelli, head of neurosurgery at the Regional Hospital in Varginha) who claims that during the 1996 Varginha crash incident he came face-to-face with a live non-human being brought to the hospital under military guard. The doctor says the entity looked unlike a human, with a large head and teardrop-shaped eyes, and that he communicated with it "not with words but through thought," which he described as extraordinarily intelligent and calm.[67]

The entity's telepathic message to physicians was that humans are "totally detached" from their spiritual selves and unaware of their true potential. Witnesses from the 1996 event (including children who saw the entity in a field) described it as weak, with reddish or large eyes and unusual head features.

James Fox has repeatedly stated in interviews that Brazilian witnesses and military sources told him U.S. military personnel became involved and ultimately took custody of the being following the Varginha incident. According to Fox, this included American presence at Brazilian military facilities and pressure to transfer materials and the entity, suggesting a covert joint operation rather than a purely domestic response. This ties in with claims of crash retrieval programs as well as those that pilot these vehicles.[68]

Chris Bledsoe's book, UFO of GOD: chronicles his profound experiences with unidentified aerial phenomena (UAP) and their transformative impact on his life. In 2007, amid personal and financial hardships, Bledsoe had a life-altering encounter near North Carolina's Cape Fear River.[69]

While fishing with his son and colleagues, he wandered away to pray for relief from his struggles, including a debilitating illness. During this moment of despair, he observed mysterious orbs and experienced approximately four hours of missing time. Upon reuniting with his group, they were collectively terrified by the sight of multiple UFOs, leading them to flee the area.

Following this encounter, Bledsoe reported a spontaneous remission of his illness, attributing his healing to the phenomena he witnessed. He began experiencing ongoing interactions with enigmatic entities, notably a benevolent female figure he refers to as "the Lady." These interactions often involved telepathic communication, where Bledsoe received messages he perceived as spiritual guidance. He describes these experiences as deeply intertwined with his faith, interpreting them as manifestations of divine presence.

Bledsoe and particularly many other UFO experiencers confirm that there is a connection between UAP encounters and spiritual experiences. He states that these phenomena may serve as conduits to the divine, challenging conventional distinctions between extraterrestrial and spiritual realms. His telepathic communications with "the Lady" have led him to believe that such entities are messengers or manifestations of a higher power, offering insights into humanity's spiritual evolution.

What should hint as significant with Bledsoe is that his book features a foreword by Jim Semivan, a former Central Intelligence Agency (CIA) officer, and an introduction by John B. Alexander, a retired U.S. Army colonel known for his research into non-lethal weapons and paranormal phenomena. They are both legends in their own right.[70]

About a year ago, on the *Engaging the Phenomenon* podcast, he was told from someone "in the know" that the UFO truth is indigestible, it is not something that humans would be able to fully grasp or entertain. It would be as if telling someone something that they couldn't do anything about. It is so great. It is so world changing that it's transformational...maybe not in a good way...we know now that atoms are basically made of nothing, we are essentially energy forms if consciousness is non-local and not being generated by our brains but we are part of a universal consciousness. How could the president tell the people this? They have thought about how they would deliver this and what they would say and they came to a conclusion that they should wait until they have a better understanding."[71]

Bledsoe's experiences have attracted attention from various government officials and agencies. He has reportedly been in contact with individuals associated with not just the CIA, but NASA (which by the way in 2025 has now been deemed an intelligence agency mirroring the CIA and is no longer a civilian space agency), the Defense Intelligence Agency (DIA), and the Pentagon.[72] These interactions have included interviews, medical examinations, and demonstrations of the phenomena he has encountered, indicating a significant level of official interest in his case.

In addition to U.S. government agencies, Bledsoe has collaborated with academic institutions and religious organizations, including the Vatican. These collaborations suggest a multidisciplinary interest in understanding the nature of his experiences, encompassing both scientific and theological perspectives.

Yes, the Vatican. Something to note in late 2025, the newly appointed head of the Vatican Observatory, Father Richard D'Souza, stated he would baptize an alien and that such beings would be "children of God". Recent allegations have been made that one of the first UFO crashes and recoveries was off the coast of Italy in 1933.[73] The observatory was established back in 1891. They have embraced science for maybe reasons yet to be revealed. They have had an interest in astrobiology and its societal implications having even been funded by NASA looking into extraterrestrial life.

Even as far back as 2008, Father José Gabriel Funes, then director of the Vatican Observatory and a scientific adviser to Pope Benedict XVI, stated that believing in "extraterrestrial brothers" is not in contrast with the Christian faith because we cannot limit God's creative freedom.

Bledsoe's case has also drawn the attention of high-ranking officials such as Timothy Taylor, who has been involved in bringing alleged extraterrestrial materials to the Bledsoe family for examination. These interactions have contributed to ongoing discussions about the potential connections between consciousness and unidentified aerial phenomena (UAP).

It looks as though humanity is being brought up to speed...at the speed of light or better put..the speed of thought. Whether guided by light within the brain or pulled across invisible planes of consciousness, remote viewing and telepathy may not just be a tool—it may be the original language of the universe and dimensions. Some of these beings may not just be up in the sky but here on earth already. Some may even be living in our forests as we will explore further in the book. What if what Jim Semivan has stated about the UFO enigma being too much to digest is because the truth is stranger than fiction?

"The universe is not only stranger than we imagine, it is stranger than we can imagine."

— J.B.S. Haldane

Consciousness, Control, and the Multidimensional Question

In a recent episode of American Alchemy, researcher Jesse Michels interviewed Kirsan Ilyumzhinov—former president of the Russian republic of Kalmykia and longtime president of the World Chess Federation—who claims he was invited aboard a non-human craft in 1997 and communicated telepathically with its occupants.[74]

According to Ilyumzhinov, the beings conveyed that they had guided multiple past human civilizations and even introduced the game of chess as a tool for teaching strategic and multidimensional thinking, while also implying that several world leaders—both past and currently active—are aware of non-human contact.

He further stated that figures such as Vladimir Putin and Donald Trump are not ignorant of the phenomenon, framing extraterrestrial presence not as speculative folklore but as an unspoken geopolitical reality discussed behind closed doors.

Ilyumzhinov says that in September 1997 he was not abducted, as Jesse kept using this term in the interview, but clarified that he was invited and taken aboard a craft by extraterrestrial beings in yellow suits, where he had telepathic communication with them about civilization and humanity. He claims they wanted to demonstrate that UFOs exist and exchange ideas at a conceptual level rather than through language.

According to Ilyumzhinov, the beings told him Earth once hosted multiple intelligent civilizations, that they themselves created the human race, and even that they invented the game of chess. He asserts that his UFO experience was acknowledged by Boris Yeltsin and later discussed with Vladimir Putin.

On the podcast he also tells Jesse that he also spoke about it with Henry Kissinger, who "seemed to know all about UFOs before" their conversation. This positions the claim as embedded within elite political awareness rather than isolated witness testimony.[75]

The interview highlights Ilyumzhinov's broader network, including close friendship with the Dalai Lama and meetings with Jimmy Carter, Pope John Paul II, and intelligence circles, implying that knowledge or curiosity about extraterrestrials spans geopolitical and spiritual spheres. Channelers akin to Bashar and others are also telling us that there is a shift coming and that they will eventually reveal their presence.

Back in 1988, something strange slipped through the cracks of our normal reality. Channeler Barbara Marciniak began receiving transmissions from beings calling themselves the Pleiadians, and what they described wasn't science fiction—it brings to mind an ancient crime scene. According to them, Earth isn't just a planet floating alone in space; it's a quarantined consciousness zone.

Roughly 300,000 years ago, they claim, a group of so-called "creator gods" intervened here, altering human DNA by compressing a once expansive, multidimensional being into something far more limited. Twelve strands down to two. But that was just the opening move.

What followed, they say, was subtler and far more effective: a frequency fence. Not a wall you can see, but one you feel. A vibrational containment field designed to keep human awareness locked into fear, survival, confusion, and division.[76] In this framework, fear isn't just an emotion, it's a resource. A broadcast signal. Media, screens, and constant noise become less about information and more about tuning forks, keeping us vibrating in a narrow bandwidth where we forget what we are.

The Pleiadians refer to themselves as the "Family of Light", claiming they are future humans returning to help dismantle this energetic cage and reactivate what's been dormant inside us all along. As the transmission bluntly states: "The ultimate tyranny isn't soldiers in the streets—it's the manipulation of consciousness so complete that the prisoners don't know they're imprisoned."

With the release of the trailer for Spielberg's upcoming film *Disclosure Day*, with a release date on June 26, 2026, several people have pointed out that in the movie poster that emerged in Times Square with what echoes an alien and its eye, there is a reflection in the eye that remarkably resembles The Pleiades, a cluster of stars also known as the "Seven Sisters" (M45). This cluster is visible to the naked eye and sits in the constellation Taurus.

You know my thing with threes, well this star system has been referenced three times in the Bible. It has been recognized as something special by many civilizations over the years. In mythology the seven sisters are the daughters of Atlas and Pleione, who were transformed into stars in order to escape the hunter known as Orion. Many people are convinced that Spielberg is one of the people that are "in the know".

Rumored to have had a screening of E.T. The Extraterrestrial at the White House when Ronald Reagan was president back in 1982. Reagan would state that *"a number of people in this room know that everything on that screen is absolutely true."* This was significant because five years later Reagan used the alien-threat metaphor in a major public speech at the United Nations General Assembly on September 21, 1987, *"perhaps we need some outside, universal threat...how quickly our differences worldwide would vanish if we were facing an alien threat from outside this world."*

Three years after the ET screening during a 1985 Cold War summit with Soviet Leader Mikhail Gorbachev, two years before his famous UN speech he alluded to the same scenario and if it would be an event that would eventually unite the two countries.

Spielberg's *Close Encounters of the 3rd Kind* had an incredible impact on me and at the time I didn't understand why it was so powerful to me. One of the most iconic scenes is the little boy opening the door and we see the orange glow from the craft. I had a similar memory that resurfaced later on during life that I walked downstairs in the middle of the night came to the door and looked out to see a UFO coming down the street, my neighbor emerged from his house as this UFO was hoving over and a white light was beamed down and he walked into it and it sucked him up. The UFO then made its way in my direction and I don't have any recall but I had the sense that it was picking me up too.

Close Encounters of the Third Kind was released in the United States on November 16, 1977.[77] A few months earlier on August 15th, the Wow! Signal was detected on August 15, 1977 by astronomer Jerry R. Ehman at Ohio State University's Big Ear radio telescope.[78] The third anomalous event to occur seemed to be connected to the first two somehow. The Southern Television "Vrillon" broadcast interruption happened a few months later on November 26, 1977 in the UK.[79] Several weeks ago, Dr. Avi Loeb has publicly made the connection that 3iAtlas may be the source of the original WOW! Signal.[80]

The UK broadcast interruption was ten days after Spielberg's movie release. Why this pattern is relevant and significant is the trailer that dropped a few days ago for his new film *Disclosure Day* which stars Emily Blunt reveals a stunning scene where her character who is in broadcast television and looks to be a weather woman when, while live on the air, her voice is interrupted and replaced with clicks and pops reminiscent of an alien language.

Here is why this is significant. In 1977 the Southern Television interruption occurred during the broader cultural transition into what many astrologers call the Age of Aquarius, a period symbolically associated with communication, technology, collective consciousness and sudden shifts in awareness.

I have a feeling that Spielberg is using the 1977 interruption as an element of the new *Disclosure Day* film.[81] Could the film be another precursor to a forthcoming communication event? At approximately 5:10 pm GMT, on November 26th, 1977, the infamous broadcast interruption occurred in southern England when the audio of a routine news bulletin on Southern Television was overridden for about six minutes by a distorted voice identifying itself as "Vrillon," a representative of the so-called **Ashtar Galactic Command."

The video stayed normal, but the mysterious voice delivered a peace-oriented warning, urging humanity to abandon weapons and evil ways, embrace wisdom, and prepare for a "great awakening" as the planet moved into a new age of peace and evolution.

A reconstructed transcript popular among enthusiasts begins with: "This is the voice of Vrillon, representative of the Ashtar Galactic Command, speaking to you... For many years you have seen us as lights in the skies..." and goes on to warn of dangers to the world while offering guidance toward harmony and evolution, though details vary in different retellings since no official recording is publicly archived. Was its message confiscated, censored or classified?

As 3iAtlas, which when you hear it as the Third Eye which is known for awakening and seeing the truth as we are heading deeper into disclosure, Space agencies' are conducting the largest planetary defense drill with 3I/ATLAS making its closest approach to earth on December 19, 2025, "the authoritative sources describe this as an observing exercise / rehearsal because interstellar comets are hard to measure precisely, *not* because 3I/ATLAS is a threat.

NASA explicitly says 3iAtlas doesn't come close to Earth enough to be a threat. Similarly, the European Space Agency calls the tracking effort simply a valuable rehearsal even though it poses no threat.[82] We have unbelievably heard over and over again that UFOs and UAPs don't pose a threat to our National Security or Air Safety, the timing isn't coincidental.

Impatiently waiting to see what the day would bring on the 19th, I woke up to a blanket of white fog hovering over the melting snow in the forest next to our property. The weather reports and statements for news stations in New England warned the public the day before that we were about to see 50-60 mph gusts of winds and potential power outages.

My wife who works in the tallest glass building in Worcester, Massachusetts told me that the day was scary and had almost a paranormal vibration. She was scared. People who worked in the office had bags packed ready to leave, they were terrified about the increasing weather as each hour passed.

Later that day over fourteen states would see from the east coast to the west coast at 3IAtlas made its closest pass to earth suffering from high winds and power outages. They weren't localized weather patterns or confined to one region or tied to a single weather system. This happened all at once. Coincidence?

I reached out to Uri Geller as I saw an article and then several popping up globally that he agreed with Dr. Avi Loeb that this wasn't a comet, but something else. With hopes of getting an interview and connecting with him on what is happening we were working on a time to connect but his timing on what he would post a on social media day before 3IAtlas made it closest path to Earth he said this, which beautifully sums up not only this chapter, but what I have been writing about for several years now that has finally manifested into the book ***Interdimensional***:

> "Friends — the interstellar object 3I/ATLAS passes closer to Earth TODAY than at any other point in its trajectory.
>
> Many scientists believe it's a vast comet; but I must be honest: I am NOT convinced that is all it is.
>
> In my view, its trajectory, timing & origin strongly suggest it may be an alien probe or interstellar **TRANSMITTOR** — intelligently launched & propelled through the universe — perhaps with a mission to EMIT an energetic frequency & consciousness changing **SIGNAL** to every civilisation it passes on its path.
>
> I am convinced it's a **CATALYST** — here to awaken something deep within us, to remind us that we are NOT alone, & that we are not merely citizens of Earth, but citizens of a vast cosmic conglomeration.
>
> Throughout history, ancient civilizations understood celestial visitors as symbols of **CHANGE.**
>
> The Romans, Babylonians, Egyptians & Mayans — almost ALL ancient cultures — saw comets & celestial events as powerful **OMENS** of **TRANSFORMATION**, knowing intuitively that the heavens & humanity are deeply connected.
>
> As above, SO below!

That is why I strongly believe **3I/ATLAS** signifies a **BIG CHANGE** for humanity & I feel this change may be linked to **DISCLOSURE!**

For decades, governments have hidden the truth about **UFOs/UAPs** & aliens. **I KNOW** this first hand!

But the public has been slowly **GROOMED** — through films, television & culture. It was never just entertainment. It was **PREPARATION!**

Spielberg's next blockbuster is just part of this final phase.

I believe a groundbreaking disclosure announcement is **IMMINENT** - one that will change how humanity sees itself & its place in the universe.

Think of 2001: A Space Odyssey. In that story, the **MONOLITH** appears at key moments to **TRIGGER** leaps in human consciousness.

I believe **3I/ATLAS** may be playing a similar role — triggering a shift in our understanding of who we are.The Universe is not Random."

Uri Geller - Facebook post December 19, 2025[83]

Chapter Eight

WICKED WEIRD WORCESTER COUNTY

Now that people are aware of me, not only as a Bigfoot researcher and Paranormal investigator but also someone who has delved into the lore and enigma that makes up the UFO/UAP story, I am also a UFO experiencer having my own encounters and sightings. I will often be approached by people who have either grown up in this area or are from Massachusetts or New England that feel comfortable sharing their story with me.

This is a testament to the fact that I have had my own experiences and since I am willing to share them during my library talks, national conference presentations or social media conversations has yielded someone sharing an experience that they can't explain, often - for the very first time. That's how traumatic these experiences can be as well as the stigma associated with them and why most keep quiet.

Due to the fact I share the strange encounters I have had that most wouldn't tell their wife of thirty years but I am telling a hundred strangers at some talk in New England, these witnesses feel that they can confide in me about their encounter or scenario with the hope of understanding that experience. I'm still learning and I don't have all the answers, but I try to be a sounding board and share relevant experiences that I'm aware of or personal experiences that resonate the same energy.

Worcester County emerges as a persistent anomalous hotspot when viewed through overlapping layers of geography, history, and infrastructure rather than isolated incidents. The region sits atop complex geomagnetic variations, is dense with natural and man-made water reservoirs known to amplify electromagnetic effects, and overlaps with documented Indigenous sacred and vision-quest landscapes, particularly around Littleton and Acton, where altered-state practices were historically performed.[84]

This same corridor later hosted the 1909 Worcester Airship sightings, some saw a bright white ball of light and others a large cylindrical shaped object pursuing or following the local commuter train across Fitchburg, Littleton and Ayer with reportedly thousands of witnesses on December 25th and the 26th, making it one of the earliest mass-witness UFO waves in New England.[85] The area continues to generate reports of UFOs, orbs, and Bigfoot encounters stretching from Leominster State Forest, which encompasses five different cities and towns and Leominster, where I was born and raised and live now in an area known as "Monsterland."

The presence of major military and research infrastructure—including Fort Devens, Natick Soldier Systems Center, and the MIT Lincoln Laboratory observatory complex in nearby Westford—adds another layer, suggesting long-term strategic or monitoring interest in the area.[86] [87] Even culturally, the region appears magnetized: *Coast to Coast AM* host George Noory spent formative years in the Fitchburg area every summer, which is a part of Leominster State Forest aka Monsterland, a connection we discussed during my appearance on GAIA's *Beyond Belief*, that the CIA and NSA were doing research or actively involved in the area.[88] It reinforces the idea that this landscape repeatedly produces experiencers, witnesses, and investigators across generations rather than by coincidence alone.

My Mom and several other children had an otherworldly encounter in Leicester. She was about eleven years of age and had a UFO sighting with at least seven or eight other children of a gigantic triangular craft hovering over a reservoir that they frequently fished in 1950. A beautiful sunny day that shockingly turned overcast by this enormous craft she said that outstretched the sky, they could see directly underneath it and the entire group ran across the street to one of their houses. She said she doesn't have any memory of what happened after that.

What is also strange about Leicester is that during the summer of 1953, it rained frogs and toads from the sky! Residents were plucking these fallen amphibians from their roofs and gutters. This has only happened a few times in history.[89] Where did they come from?

I find it interesting that this is the location of several compelling reported UFO sightings over the years. Leicester is the next town over from Worcester where this past December I witnessed a rainbow colored UFO hovering near the town of Boylston. I first noticed it at my house in Leominster. When I arrived in Worcester to meet my wife at an event she was working at I drove onto Main Street, all the street lights, building lights were flicking all the way down the road. I even posted it on social media which garnered millions of views.

Long-time Worcester residents that I spoke to personally said they have never seen anything comparable to this before. This was happening in other states as well directly after a UFO or drone sighting. One was in Tennessee at a Tesla dealership and another was in Florida at an apartment complex if I can remember correctly. Both of these other videos were mere days apart from my video I had posted.

A few days later after the post went viral on Facebook and Instagram with over 4 million views and counting, I received an email message from a gentleman who goes by the name of "Brad" who now resides in Arkansas but grew up in Leicester, Massachusetts. He has asked me to keep him anonymous so Brad is a pseudonym. Leicester has a place in my family's timeline because my mom grew up there and I spent much time there at my grandmother's and aunt's house just behind it when

I was a kid. So I immediately perked up when I was reading his message.

Brad saw the video and reached out with the initial email. I arranged a call so that I could learn more about Brad's experience. Between the hours of 1-3 AM, Brad and his sister would hear a low humming sound that seemed to emanate from the abyss. This hum almost seemed to occur every night. He had visions of floating outside of his room and into his sister's room where he would then exit floating outside the window!

There was one time that he remembers being terrified because they heard something running on the roof of the house. Brad's sister watched a UFO for twenty minutes from her room in the same point in the sky where he would always be magnetically drawn to. Brad also remembers a time when he and his buddies were watching "satellites" and one without warning stopped. They freaked out.

Back when Brad was about eleven or twelve years old, which would put this event into the year of 1978 or 1979, when on Christmas Day he had an experience that was so vivid that he just couldn't shake it. He knows for a fact that this was not a dream but something that actually happened. Recently retired, Brad enlisted in the Air Force for twenty-seven years, serving in Operation Desert Storm and Desert Shield working on C130s and A10s. Not an intelligence operative or special forces. He was an aircraft mechanic by trade.

Ronny: "So Brad, did you ever see anything strange about serving overseas?"

Brad: "That's the funny part. One of my motivations of joining the Air Force was the hope that maybe I would see something...but nothing. Absolutely nothing."

We both laugh.

Ronny: “As a kid, do you remember or have any recall of seeing any strange lights or a craft in Leicester?”

Brad: “This is going to sound really weird, but as a kid when I was outside playing I always had the sense I was being watched by one section of the sky. I just had the heebie-jeebies feeling that something was there when I looked at that part of the sky.”

Ronny: “So you could sense it but you just couldn’t see it?

Brad: “Even as a kid I always was extremely uncomfortable with that part of the sky. It was more Southwest than where we lived.”

Ronny: “So why I am even more curious about your experience is that this is the same city that my Mom grew up in. She had a UFO sighting of a triangular craft so large it blocked out the sky hovering over the local reservoir. All the neighborhood kids saw it. They all went running into the neighbors house but that is all she remembers of that incident. She doesn’t have any memory of what transpired after that.”

Brad: “Do you remember when Whitley Streiber wrote that book Communion?”

Ronny: “I was literally thinking of him and that book specifically before I called you.”

Brad: “The only reason I say that. When that book came out I was already in the service back then and my mom was a devout Catholic, we were raised Catholic. Around 1130PM, my mother, sister, and my brother-in-law are coming back from my other brother’s house. They were descending down Forest Street when without warning there was this big white bright light down at the end of the road over the power lines.

It wasn’t a helicopter and there was no noise whatsoever and they all watched this thing come up the road.This was enough to freak my Mom out that she exited the car and ran into the house screaming. His sister runs in after her and his brother-in-law just continues to watch the object. My sister actually reported this sighting to the Leicester Police and they sent a car out"

Ronny: "What year was this?"

Brad: "This was the same night that Whitley Streiber claimed to have an event in upstate New York just a mere few hours later. His incident was between 100-300AM the same morning.

(Author's note: This is about a four-hour drive from Leicester)

On another note, my best friend along with his wife who lived in Auburn, a few towns over, witnessed a giant craft hovering over the old folks home. It was completely silent. He was in the Air National Guard and knew what types of aircraft we had in our arsenal. He was literally underneath it and couldn't hear a sound. This occurred when I was stationed overseas as well."

What I find extremely interesting is that some first-time experiencers or long-time alien abductees have mentioned that when they have evoked the name of Jesus Chris or started praying during these frightful and traumatic events the strange activity will essentially stop immediately in that very moment. It has led certain researchers to believe that what we are dealing with is something evil or demonic.

Brad: "We always heard a hum. We thought it may just be Worcester Airport since it was twelve miles away Maybe we were just hearing the engines. This was a distinct hum. We lived in a small cape with two bedrooms on the top floor. My sister's bedroom was upstairs and I shared the other room with my older brother who was ten years older than us.

I was lying in bed and woke up and heard the hum. The weird part is when I rolled over to look towards the middle of the room and faced my digital clock that had these blue LED lights. At that moment, something touched my left temple and something flashed. Remember the old polaroid flash cubes? Something flashed. It wasn't white it was resembling an orange phosphorus - I say phosphorus because it crackled like a firework dissipating. It was quick and so intense, like boom, instantly I fell back in my bed.

My head went back towards the wall, as though I wasn't allowed to see anything in the middle of the room. You know how when your arm falls asleep? You know what it feels like?"

Ronny: "Yeah, that tingling sensation"

Brad: "It started at the top of my head and went straight down my entire body all the way to my toes. I was thinking, "What the hell is going on?' I was freaking out. I tried to yell for my brother. He was in the other bed and nothing was coming out. The only thing I knew what to do was close my eyes and start repeating the Lord's prayer over and over again."

Ronny: "Is that something that you had learned before?"

Brad: "I was a devout Catholic. I was an altar boy for fifteen years until I left for the service when I was eighteen. But it was the only thing in the moment in time that I knew what to do. How am I getting out of this? So I started to pray.

I remember I tried to look down towards the end of my bed. I don't remember seeing anything, anything at all. All I knew was that I couldn't move. I was maybe 30 to 40 seconds into these prayers when the humming stopped. The hum was gone. As soon as it disappeared, just the reverse where the tingling stopped from the bottom of my feet and all the way up to my head and I got control again.

I don't know if you have heard of anyone experiencing something like this before? It was weird. Since that event, I have slept in alert status constantly. Between 1 and 4 AM I don't really sleep."

Ronny: "Yes - that sounds very traumatic and it's now ingrained into your DNA. The fear. Wow. Do you remember seeing or experiencing anything out in the woods or forest where you lived?

Brad: "I used to enjoy shooting with my .22. Sometimes I would go by myself. One time I saw this big old owl and it was just sitting there looking at me. It was huge. It was 2 to 3 feet tall. It seemed very odd. I couldn't even tell you what kind of owl it was.

Ronny: Tell me about the time that you and your sister heard something running on the roof? Was that the same night of your paralysis experience?

Brad: "I have had that one experience of something running on the roof. We would hear things running across the roof. We had an attic door on my sister's side. We would hear stuff moving around in there, a lot. To the point that we called our Dad to check it out and he couldn't find any evidence of animals, like feces or mouse droppings. Nothing. I have other thoughts, whether they were real or just dreams I don't want to speculate."

Ronny: "What are some of those things that you think might be dreams? I know you may be hesitant but there are interesting things that I may recognize from other cases or conversations. Could you share one of those without fear or judgement?"

Brad: When I was growing up as a kid, my brothers and sister had their rooms up on the second floor. I was in a crib in my parent's room facing the hallway for the first couple of years.

I think I have a memory of maybe looking off that crib and down that hallway one time and seeing...seeing a few things walk down that hallway and out the door. I can't tell you if...

Ronny: " Like, gray aliens?"

Brad: "Yeah, I didn't want to go there but I can't tell you for sure if that something is true or just in my head."

Ronny: "It could be one of your first memories too as a kid."

Brad: "I wasn't scared. To me, it was as if it was someone to play with or something to that effect. I was probably four." As a person, I have always been very, very alert. People think I am crazy but I can remember a lot of things. I can remember my Dad holding me when I was born. I remember coming home from the hospital. I have that type of memory.

I think there is a lot more to us than we know."

Ronny: "Agreed. I recently learned that this area and specifically the towns of Acton and Littleton were ground zero for the tribes from New England to come and do their vision quests. There is something special about this area. Whether there is an opening or a portal or something. Why would there be more activity here in comparison to other locations? Bigfoot, UFO stuff, all kinds of weird stuff. What about your kids?"

Brad: "My kids have seen something in the house here in Arkansas. They claim to have seen a blue orb or light in the house."

Ronny: "Oh wow - that's interesting. What about your wife? Any experiences as a kid? Has she had any weird experiences?

Brad: We think we see things sometimes here in Arkansas by the corner of our eyes, in our peripheral vision, but nothing definitive. As a kid, she had one experience and it wasn't as intense as mine. She was staying with her father down in

Dallas. She said she could feel something breathing over her and into her ear when she was nine or ten years old. This freaked her out. She yelled and ran into her father's room.

I will tell you this. I told her about a dream that I had when I was a kid. She told me she had the exact same dream."

Ronny: "Oh my God, tell me."

Brad: "My wife and I have been married for over thirty-six years. When I was kid, I had a lot of dreams about living in that neighborhood. To this day, I still dream about this neighborhood of Leicester. All I can say and the only way that I can express it was...in this dream, I looked up into the sky and what I saw was manhole covers...everywhere. Now my wife told me that she had this exact same dream."

Ronny: "Do you think that you were seeing a future event?"

Brad: "I don't know. Some of the future events I have seen are not good. So, I hope not. But the experience I shared with you was real. This was real. It happened on Christmas Day, the same evening that Whitney had his experience"

Ronny: "Whoa - wait a minute. This happened on Christmas? Are you familiar or aware of the Worcester Air Ship sightings of 1909?"

Brad: "No, I think I heard about it."

Ronny: "Okay - so what is wild. Is that in 1909, hundreds of people across cities in Ayer, Fitchburg and Worcester witnessed this gigantic UFO? Some people only saw a bright light. It just happened that this event lasted a few nights...***December 25th and 26th***! 70 years later."

Brad: "Really? Now that's weird."

Now a few things to reflect on. What are the odds that they both had the same dream as children? Does this confirm their destiny of sorts? Aligning other abductees or experiencers who are miles apart. He met his wife on his first assignment in Arkansas. He selected this location because of the Bigfoot film "The Beast of Boggy Creek". He has lived here most of his life. He comes back to the east coast to vacation up in Maine in the summertime.

Clearly his wife is connected and they are meant to be together. She has had experiences as a child as well. Brad told me of his kids, one of his daughters would be considered sensitive or a medium of sorts. She can communicate with the spirit world. The kids proclaimed to see a blue orb together inside of their house. Abductees or experiencers will often have gained some sort of "super power" after their experiences. Psychic abilities let's say. Activating the senses and long lost abilities that the ancients once had. Brad talks about having an uncanny ability to remember past events with clarity and capacity that is unlike a normal human being in my opinion.

While interviewing him I also shared that I have had sleep paralysis and night terrors as a child. I always believed that there was someone in my room. I don't remember the age but I might have been six or seven and I can recall seeing small golf ball-sized balls of orange fire floating from behind me and I could see them in my peripheral vision on the right. When I would quickly turn my head they would fly out of view.

My wife witnessed one of these in our home when we were first married and after casting a Bigfoot print in Leominster State Forest in 2010. One of my daughters also had one in her room hovering and lighting up the darkness of the room. My Mom had decided that we were going to leave the Catholic Church, as we were attending St. Cecilia's church where a lot of the French Canadians that migrated down from New Brunswick, my father being one of them at the age of four, lived near the church in an area known as French Hill.

My Mom decided that we would be attending Leominster Assemblies of God and our faith was shifting from Catholic to Pentecostal. I would have to say that this is when strange things began to happen when we eventually moved to Union Street from French Hill. This was at the bottom of Elm Street where I currently live which is only a mile and a half away from where I grew up. We are still living in the zone of all this strange activity and less than a mile away from where the Bigfoot prints were cast and sightings of UFOs and Bigfoot continue to this day.

Are there significant dates that seem to evoke these craft? Are they always here and just show themselves during these specific periods in time? How is it that such a significant UFO event in central Massachusetts history occurred the same time and year of his Leicester experience and that of Whitley's in upstate New York just a few short hours later?

Why are certain dates or numbers that have this connection seem to appear along with the same locations or distinct locations? Almost as if the combination of these potentially unlocks the key for a visual sighting or experience for those who are lucky or rather unlucky.

The Rendlesham Forest incident, often referred to as "Britain's Roswell," took place in December 1980 near RAF Bentwaters and RAF Woodbridge, two military bases in Suffolk, England. Over the course of several nights, U.S. Air Force personnel stationed at these bases reported witnessing strange lights in the nearby Rendlesham Forest.[90]

One of the most notable sightings occurred on the night of December 26th, when servicemen including Deputy Base Commander Lt. Col. Charles Halt observed a series of unexplainable lights that appeared to hover and move erratically through the trees. The witnesses described triangular shapes, flashing lights, and even possible landing marks, sparking widespread interest and debate about the nature of the event.

Just a few days after Rendlesham, on December 29, 1980, the Cash-Landrum incident occurred in Texas. Betty Cash, Vickie Landrum, and Colby Landrum were driving near Dayton, Texas, when they encountered a diamond-shaped object emitting intense heat and bright light. According to their accounts, the object was surrounded by a formation of military helicopters.

The witnesses suffered severe physical symptoms after the encounter, including burns, nausea, and radiation-like effects, prompting them to seek medical treatment. The Cash-Landrum case remains one of the most compelling UFO incidents in the United States, largely due to the documented physical effects on the witnesses and the possible involvement of military aircraft. What makes these incidents particularly intriguing is their proximity in time, with the Rendlesham event occurring just three days before Cash-Landrum.[91]

Both incidents involved unexplained aerial phenomena and potential military involvement, though the nature of the objects observed and their effects on witnesses differed. The Rendlesham incident focused on strange lights and potential extraterrestrial contact, while Cash-Landrum involved a direct physical encounter with a UFO dripping some sort of substance or after material.

The timing and military connections in both cases have fueled speculation about a possible link, suggesting that there may have been increased UFO activity or secret military experiments occurring in late 1980. We know that UFO events over bases typically pointed to the fact that these bases had nuclear munitions or capabilities that the craft seemed interested in.

Both cases involved multiple witnesses and reports of crafts leaving behind physical effects—radiation burns in Cash-Landrum and traces on the ground in Rendlesham. The

December connection points to possible UFO activity during the holiday season.

Now that we know we have recovered UFO technology and or craft along with biologics points to the possibility that some of these bases might have been hiding this technology underground and underneath some of these bases, hence the attention from the UFOs. The nuclear aspect is essentially the cover for the crash retrieval programs.

When looking at UFO events over the past hundred years, July tends to stand out as one of the most active months for sightings. Several major incidents have occurred in this time frame, including the famous Roswell incident in July 1947, which remains one of the most well-known UFO cases in history. Additionally, the Kenneth Arnold sighting, which helped popularize the term "flying saucer," also took place in June, but it was in late June leading into July that UFO interest began to spike, particularly in the late 1940s.[92]

Another significant period is often early fall, especially September. Many notable sightings have occurred during this time, including the 1961 Betty and Barney Hill abduction, which happened in September.

While specific weeks can vary by year, there evidently is a recurring pattern of heightened activity in both summer and early fall, suggesting that these might be key periods for UFO sightings historically. Theories for this spike range from more people being outdoors during summer months, to possible atmospheric conditions that might contribute to sightings, or even patterns in alleged UFO activity itself.

I also wanted to point out something else that I noticed when comparing locations in Massachusetts to those in England after my trip there. I noticed an interesting pattern that I was previously unaware of.

Why is it that cities or towns that have the same names have the same strange phenomena thousands of miles away? A lot of the towns and cities in Massachusetts are named from cities or towns in England. These same cities and towns have very similar activity and sightings from orbs to cryptids to UFOs and time slips. Is it just a coincidence or is there something to it?

Are these all simply coincidental patterns or something else that would help us determine the time and place of these future appearances?

Chapter Nine

MYSTERY NEW JERSEY DRONES AND EAST COAST UAPS

In November of 2024 something started to shift in New Jersey. Residents in Morris and Somerset counties reported observing large drones, approximately six feet in diameter, flying at night between dusk and 11PM. On November 19th, The Morris County Prosecutor's Office confirmed that law enforcement officers on patrol had observed drone activity the previous night.

Three days later, on November 22nd, the Federal Aviation Administration (FAA) issued a temporary flight restriction over President-elect Donald Trump's Bedminster golf course, effective through December 6, 2024.[93] Another three days went by when the FAA would impose another temporary flight restriction over the Picatinny Arsenal military base in Morris County, lasting until December 26, 2024.[94] The following day, the FBI's Newark Office announced collaboration with state and local partners to investigate the ongoing reported mysterious drone activity.[95]

New Jersey Residents near the Boonton Reservoir in Morris County reported seeing multiple drones over consecutive nights on December 1st and December 2nd. The very next day, on December 3rd, the Morris County Prosecutor's Office stated that the drone sightings posed "no known threat to public safety." Wow, that was fast! Two days later, on December 5th, Governor Phil Murphy held a briefing with federal and state officials to discuss the ongoing drone sightings, emphasizing active monitoring and coordination. Just two days later on December 7th the mysterious drone sightings began to expand in location and in numbers of drones being witnessed. Drone sightings would then begin to occur in at least eight counties in North and Central New Jersey, including Bergen, Essex, Hunterdon, Morris, Passaic, Somerset, Sussex, and Warren.

The very next day, a group of Morris County mayors and New Jersey lawmakers sent letters to state and federal officials, urging for a thorough investigation and more information regarding the drone activities! December 9th, Governor Murphy addressed the media again, expressing serious concern about the drone sightings and acknowledging the sophistication of the drones, which could "go dark" once spotted.[96]

During the height of these drone incursions in New Jersey, I reached out to a fellow colleague Scott Davis who is a police officer within one of these townships where they are having these drone sightings. He also happens to be a psychic and has done a lot of remote viewing work. I asked him if he would look into these drones for me via remote viewing. He complied. He said that what he was experiencing was very odd. "It is as if I am encountering an electric fence."

He described seeing what others are reporting which are not only car-sized drones but also these orange and reddish balls of fire or orbs. He said that when he looked into them he witnessed an object that resembled an orange orb and he was being restricted mentally as well as physically trying to access its origin or intent. He was restricted from seeing.

He said that this was the first time that this has ever happened to him. I asked him because this was the first time he had experienced this if he believed that he might be interacting with someone or something that is of non-human origin or intelligence? He said, "Yes, I think so."

Across the pond in Manchester, United Kingdom, in late November 2024, a commercial co-pilot shared a video and several photos which quickly went viral. The video is of a metallic looking sphere hovering high above the tarmac as it looks like the plane is awaiting takeoff on the runway during a misty, greyish day. There are subsequent photos of a blue-metallic looking sphere with other elements on its exterior and based on a shadow below the object, which looks as if it is only inches or feet above the tarmac! Shared on X the object would then disappear in a straight shot movement up into the atmosphere as airport security approached the object.

There has been some dispute on Reddit and other channels about the authenticity of the video and images as the user account was quickly shut down on X after the video and story went viral. This happens a lot when someone posts something significant and then they don't want the negative attention or repercussions from their employer for sharing. There has also been discussion based on the weather that day and controls in the photos that this was an image taken in June of 2024 and not November. This affects the images and its owner's credibility. Its timing of it being shared prior to the activity nonetheless is intriguing and shows that it wasn't looking for the attention during the wave but prior to it. It could be a legitimate image of what people are seeing all over the world.

With all the reported drone sightings over military bases and those bases being shut down because of these "drones" or strange orb-shaped lights and objects throughout the United States, Japan, United Kingdom and Australia to name a few, it could be a piece of the puzzle we are trying to solve.

On December 11th, 2024, the Pentagon gave a press conference updating the latest on the Ukraine and Russian war and a question was quickly asked about the "Mysterious Drones" that were appearing over New Jersey. "Our initial assessment is that these are not drones associated with any foreign entity or a foreign adversary. We will continue to monitor what is happening. These are NOT US military drones and they are not sanctioned by the FAA."[97]

One of the theories that was posed during the press conference in DC from one of the many reports on their origin is that a New Jersey congressman Jeff Van Drew says he knows where the drones are coming from. He states that an Iranian warship was launched a month ago and is sitting somewhere along the east coast in the Atlantic and these drones are coming from that ship. This was deemed by the Pentagon in a following press conference as false and not the source of these drones.

We then learned that these drones were being seen all along the east coast and now within the borders of New York. On December 14th, New York Governor Kathy Hochul announced that the federal government was deploying a drone detection system to the state to assist in investigations, following similar sightings in New Jersey.[98] National Security Council spokesperson John Kirby said in another press conference on December 12th that most of these drones being seen over New Jersey and the east coast are manned aircraft and they have not been seen over restricted air space as previously reported. There has been no determination that these are a national security or safety threat to the public.[99]

Don't worry about a thing, New Jersey. The FBI and Homeland Security are investigating this with local authorities and it is ongoing but this is in their initial assessment, thus far, that there is nothing to worry about. They cannot characterize what these car-sized drones are, where they are coming from and where they are going but they are deemed not a threat.

Shouldn't you consider these anomalies a threat first and then deem them not a threat afterwards? If they are not ours, they are not from a foreign adversary, where the hell are they coming from? The ocean.

And when authorities tell us that these are mostly legal, manned drones being misreported and misidentified by the public, you have some solid lawmakers coming forward with their own videos and sightings of these drones! Former Maryland Governor Larry Hogan shot video of several of these drones from his hometown in Davidson which is only twenty-five miles from the United States Capitol. No concern there.

New Jersey Senator Andy Kim shared his video of what looks to be two drones flying towards each other. The Senator then checked a flight app and was unable to locate the two objects he was filming, determining that these were not manned or legal planes hovering in the sky. They were something else and potentially from somewhere else.

Oh but there is more. Reports of these drones were seen hovering over La Guardia airport! Since November 18th, the FBI had received over 3,000 reports or tips regarding the drone sightings with an average of about 92 drones sighted per night...all over New Jersey! Equivalent to the legend of the Men in Black, I was hearing rumors online and from others that some of the videos that locals were capturing and then posting on Instagram and TikTok were then being paid visits..at their home by individuals telling them to take down their drone videos!

We are just getting started, but let's rewind a bit here back to 2023, December 2023 to be exact, when for seventeen consecutive nights a swarm of drones plagued Langley Air Force base in Virginia. For over two weeks these drones flew over restricted airspace but they were not shot down. Had they tried before and failed? Due to security concerns, the F-22s that were situated at the base were moved to another location! These drone excursions would begin about forty-five minutes after sundown and they were observed by military personnel on the base.[100]

The purpose and *origin* of the drones was still under investigation by authorities. They didn't know if these were hobbyists, commercial drones or some sort of foreign military intelligence collection campaign. In 2019, dozens of these drones, what we now know are tic-tac UFOs, were seen shadowing Naval warships for weeks along the California coast.[101] Since that time there have been dozens of incursions over military and civilian installations including the Palo Verde Nuclear plant in Arizona, the secretive Plant 42 in Southern California where defense contractors were building the newest stealth fighters (more than likely the F-45s which Boeing just won the contract for in March of 2025)

Back to New Jersey where News Nation field reporter and local resident Rick McHugh investigated the drones in Red Bank. As a New Jersey resident himself, he didn't expect much and thought that he would find out that this whole drone thing was due to a bunch of hoaxers and would have this all wrapped up in a few days.

McHugh humbly stated "What I saw was more sophisticated than I could ever imagine. We have been out here for about an hour and we have seen forty to fifty of these drones. If you look behind me you can see one over my shoulder right now." He then turns his back and we see this enormous drone on live television hovering in the background. The faint hum of what sounds similar to propellers rotating in the distance. He goes on, "These drones are about eight to ten feet wide, colorful with blinking white lights."

Hours earlier he was invited to join the Ocean County Sheriff's department as they launched their own military-grade drones to see if they could intercept the mysterious drones residents were spotting in the sky. Soberly, they too were faced with the strange reality. "These drones are able to avoid detection because they don't give off heat like regular drones." McHugh went on. He then describes that one of the Sheriff's witnessed fifty of these drones come off of the Atlantic ocean heading towards the mainland of New Jersey. The officer immediately called 911! [102]

The United States Coast guard was then dispatched to investigate and they encountered thirteen of these drones following their ship! "The Ocean County sheriff's office has a drone unit and while using an industrial drone as they approached one of these craft it quickly evaded them and disappeared out of thin air."

The drones are capable of avoiding and evading radar, electronic detection as well as reports of police helicopters when in pursuit. The use of drone guns to disable these drones are also not working! These drones are going up and down in elevation, moving erratically and seen only hundreds of feet above homes. A helicopter that was above one of these drones aborted as he feared that his machine would be in some sort of jeopardy.

Some report these drones being silent and others with that propeller sound. Aforementioned, residents are seeing these strange balls of light as well, separate from these drones, but these objects are all being labeled under the "drone" category which I find interesting.

As of April 2025, we still don't know what they are, where they are coming from or who is operating them. There were rumors that there was a dirty bomb loose somewhere along the east coast and that these were advanced nuclear-sniffing drones looking for that bomb. The problem with this theory...worldwide sightings of these same craft during the same time period. The New York governor revealed during one of his briefings with Homeland Security and the FBI about the drones that shared with him that when we put a spotlight on these drones...they go dark. Does that mean that they disappear?

During an interview with Ross Coulhart on News Nation, Mayor Sam Morris of Mine Hill Township, New Jersey shares his thoughts about the "drones". "I don't know why they are going dark when there is a spotlight placed on these objects. They are not making it a secret that they are up there and that they don't want to be followed. They come out at 530PM when the sun has set and they stick around until about 1030PM or

11PM and then...disappear. They are incredibly obvious and they are not trying to hide. If they wanted to be hidden they would come out at 2AM, they would go through the sky dark and we wouldn't even notice. We wouldn't even know. I don't even know why they are doing that." Morris said.

The drones started appearing over Morris County and then eventually spread to all the other counties. Every single one. The surprising thing is that the number of sightings over Morris County didn't go down when the sightings expanded. The other counties were reporting the same amount of sightings which means that their numbers, the amount of drones people were seeing was expanding exponentially between all of the New Jersey counties. There were 18 reports, 11 confirmed sightings of these over and in restricted military airspace over Picatinny Arsenal with the first one being reported on November 13th, 2024.

After weeks of these drone incursions Ross Coulhart from News Nation shares that he has been receiving reports of UAPs, drones showing up all over the world hourly and almost minute by minute. People are seeing glowing luminous orbs, orange orbs and triangle shaped blue colored lights. A majority of these sightings are seen over United Kingdom sensitive military bases such as RAF Feltwell and RAF Lakenheath where it is believed that the United States is hiding portions of their nuclear arsenal. Intelligence officials within his circle have shared with him that these are not aliens, UAPs or UFOs but adversarial craft from enemy countries like China or Russia.

One of theories of why the US hasn't revealed the source of these drones is because if we were to deem that they were Chinese or Russian, that declaration would be essentially announcing an act of war due to the fact that these drones are flying over our sensitive nuclear and military bases. Most of our nuclear power plants are located in New Jersey and along the east coast so this theory isn't that farfetched.

However, many of the military, police and other witnesses of these mysterious drones state that some of them are seeing orbs or balls of light and other anomalous lights in the sky, not drones. Could these drones be aware of the emergence of these *other craft* and they are being used as a sort of cover up for the UAP and UFO sightings? There are 21 counties in New Jersey now there have been sightings in all of these counties!

A few months prior during September 2024, the Canadian government released an image of the UFO or UAP that was shot down by US Military over Canadian soil back in February of 2023. If you recall, for several weeks, North America was being inundated with "Chinese Spy Balloons". A strange-looking object indeed and strange timing to release this image as it nicely fit into all of these recent drone incursions. The object in the photo (see Figure 9.1) was seen hovering in the sky and it was eventually taken down in Canada's Yukon territory near the Alaskan border.[103] [104]

The North American Aerospace Defense Command first detected the "high-altitude airborne object", car-sized, at around 40,000 feet over Alaska. Jets were quickly dispatched to observe it. The object moved from Alaska into Canadian airspace before being downed on February 11th, 2023. This event followed the downing of a Chinese surveillance balloon and was one of three aerial objects shot down in that month.

Now less than a week before a Chinese spy balloon was shot down along the South Carolina coast. F-35 Pilots that went up to shoot down the Alaskan anomaly had differing reports. One of the pilots discussed that the object "interfered with their sensors" on the planes, but not all pilots reported experiencing this phenomena. Some of the pilots also claimed to have seen no identifiable propulsion system. They could not explain how it was staying suspended in place in the air, despite the object cruising reportedly at an altitude of 40,000 to 60,000 feet.

Here is the timeline of what occurred leading to the shootdowns of these objects.

• **January 28 – February 4, 2023:** A high-altitude Chinese surveillance balloon traversed U.S. and Canadian airspace before being downed off South Carolina's coast by a U.S. Air Force F-22.

• **February 2, 2023:** Another Chinese balloon was observed over Latin America, passing over countries including Costa Rica and Colombia.

• **February 10, 2023:** A high-altitude object was detected and shot down over Alaska's Beaufort Sea by the U.S. Air Force.

• **February 11, 2023:** An unidentified cylindrical object violated Canadian airspace over Yukon and was shot down by a U.S. F-22 on orders from Prime Minister Justin Trudeau.

• **February 12, 2023:** An octagonal object with hanging strings was tracked over Montana, Wisconsin, and Michigan before being shot down over Lake Huron by a U.S. Air National Guard F-16.[105]

Initially, authorities were confident that a recovery mission would be successful as this shootdown had occurred over the ice in the Yukon. Conveniently, while the United States and Canada initially searched for debris from the object shot down over the Yukon in February 2023, reports indicate that the search was called off due to weather and the low probability of recovery, and no debris was recovered. Or did they find something that they didn't want the rest of us to know about?

It's been a long time coming but various researchers, self-proclaimed government insiders and deep state whistleblowers warn that we are about to see a staged alien invasion. An event that will either unite or create chaos. Have events from the past been averted into alternate realities to avoid our demise or is this just another example of fear being used to control the masses? It sure does feel as if we are entering into a new phase of humanity. Or is it just another illusion?

A few weeks ago I received a text from Scott Davis, New Jersey police officer, with a frantic message. "There is a freaking drone over my house right now! This hasn't stopped dude. They are still showing up!"

THE SPACE TRANSIENTS OF THE 1950s

Astronomer Dr. Beatriz Villarroel with the Nordic Institute for Theoretical Physics recently published a paper which is under peer review regarding the possibility of potential alien craft in geosynchronous orbit during the 1952 UFO flap over Washington D.C and a potential correlation of these craft emerging and appearing after we began detonating the atomic bomb and nuclear activity.[106]

Why this is so significant and mind-blowing is that this time period of analysis is pre-satellite, so these objects were not created by man as we know it. We didn't launch satellites into space until 1957.[107] We were not at this technological level at this point in time. Currently, and this is something to ponder, 70 percent of these objects in our orbit are still unidentified in 2025. Could these be aliens or an earlier human civilization that had achieved this great feat? Beatriz states that the reason for these transients, these glints of light, hinting that it is sunlight reflecting off of the surface of some object. These are potentially in our orbit for...surveillance. These are technosignatures that have been researched for years.

The primary telescope at Mount Palomar Observatory is called the Hale Telescope. It is also known as the Palomar 200-inch telescope or the 5.1 meter Hale Telescope. The telescope was dedicated in 1948 and named in honor of George Ellery Hale, whose vision and leadership were instrumental in its creation. The sky survey glass plates (1942 to 1957) that were used in the early 1950's have been digitized and Dr. Villarroel and scientists around the world have been researching these "transient" objects. There are hundreds of thousands of them that appeared on photographic plates and then disappeared. These objects are considered to be technological, constructed, artificial and not natural.

Statistical analysis (including Earth-shadow testing) suggests that these are highly unlikely to be mere photographic artifacts which would indicate that these are actual objects reflecting sunlight. What is intriguing is the timing of the most significant transient candidates coincides directly with the week of the 1952 UFO flap over Washington DC! This was probably the most significant sighting as radar operators, pilots and thousands of others reported unexplained and unidentified flying objects in our skies.

These transients also coincide with the nuclear bomb tests![108] The modern UFO enigma seemed to emerge after the Roswell crash of 1947, three months later the CIA was created and modern day experiencers and abductees recounted the message they received from the others that what we do in our world affects their world so they have a vested interest in our activities.

Dr. Villarroel decided to look at transients that were falling within the shadow cone of the earth and to her surprise they were there! What they didn't realize was that when they looked at the transients being visible and vanishing and occurring on the same nights - July 19 (triple transient case) and July 27th (the top candidate with a -0.0001 probability of being chance). It is extremely unlikely by chance that there would be this alignment. It could be a fleet of ships if they are talking about several hundred thousand transients with 30 percent of them being potential craft!

These unidentified objects, pointed-like lights, perfectly aligned which then disappeared within a single exposure. The time between two plates is about 30 minutes. These flashes or reflections are specular reflections as clear as mirrors. She estimates that 30 percent of these transients are specular objects. We see these now as there are many satellites and space junk in orbit. But back during this time period we were still five years away before we entered into space. These are not ours. I wonder if we would see something similar in 1947 which would coincide with the Roswell crash among other significant plates and what secrets they may hold.

This highlights a potential link between astronomical anomalies and UFO sightings. The statistical link showing more transients near nuclear test periods also suggests a possible causal mechanism - maybe unintended electromagnetic effects, reflections or even deliberate orbital activity. This is not the profile of random surveillance technology or mass misidentification, but something far closer to what Jacques Vallée once described as a control system, a phenomenon that shapes belief, response, and attention rather than engaging in overt confrontation.The pattern suggests intelligence operating across layers of perception, not merely airspace.[109]

This is not the first time anomalous aerial phenomena have appeared near sensitive military assets and then dissolved into official ambiguity. Similar waves have unfolded repeatedly across decades, often accompanied by contradictory explanations and the quiet withdrawal of definitive answers. As J. Allen Hynek ultimately acknowledged after years inside *Project Blue Book*, the persistence of these encounters points less toward misidentification and more toward a phenomenon that adapts to human expectation and technological context.[110] [111]

Donald Menzel, an astronomer, was a prominent figure in the Project Blue Book investigation of UFOs, specifically as a scientific advisor and a skeptic. We know now that Project Blue Book was the Air Force's way to calm down the public and used to discredit UFO witnesses and mitigate the pandemonium. Menzel authored books for example "The World of Flying

Saucers" and "The Moon, Meteorites, and Comets" where he analyzed UFO reports, often attributing them to natural phenomena such as atmospheric distortions or misinterpretations of celestial objects. Menzel testified before Congress, stating that all UFO sightings had natural explanations like temperature inversion or misidentification.[112]

A few months later after the DC UFO flap in September of 1952, Menzel became the acting Director of the Harvard Observatory. Under his leadership, around one-third of Harvard's sky-survey plates were destroyed with survey work suspended for another 15 years! Can you say cover-up!? What was on those plates! His actions coincided with the suppression of potential corroborating data. Were they destroyed or does some agency or group have possession of them?

I think we have a good idea of what they were hiding. It was discovered in the 1980s when researchers had gained access to his personal files that Menzel had a documented relationship with the CIA and NSA. He was involved in classified work in cryptography and signals intelligence for the US Government and had a top secret clearance and consulted for several intelligence agencies. His role is believed by many to have been to dissuade and debunk the UFO enigma.

Dr. Villarroel says that since this discovery she didn't sleep for several days and now is confident that we are not alone...and that we may have never been alone. She expects pushback based on the current stigma and isn't confident that her peers will push the paper forward. Could these objects still be there now?

From the 1952 Washington D.C. flap to the orbital transients of the early Cold War, and now to the drone-and-orb wave over New Jersey, as well as the east coast, the signal remains remarkably consistent: emergence near nuclear activity, strategic silence from authorities, and phenomena that straddle physical craft and luminous, non-mechanical forms. These are not isolated incidents separated by decades, but recurring expressions of the same underlying presence adapting to each era's technological and cultural filters. If this is an intelligence, it does not announce itself—it reveals itself, incrementally, where land, sea, air, and space converge.

John Keel referred to this spectrum of appearances as operating beyond simple physical categorization, suggesting an intelligence that intersects with our reality rather than arriving from it. If these events are connected, then what we are witnessing is not escalation, but continuity.[113] Remember coincidences don't exist and what makes this wave different is not scale, but the timing, the repetition and the proximity to thresholds humanity has crossed before.

Buckle up it is going to get weirder.

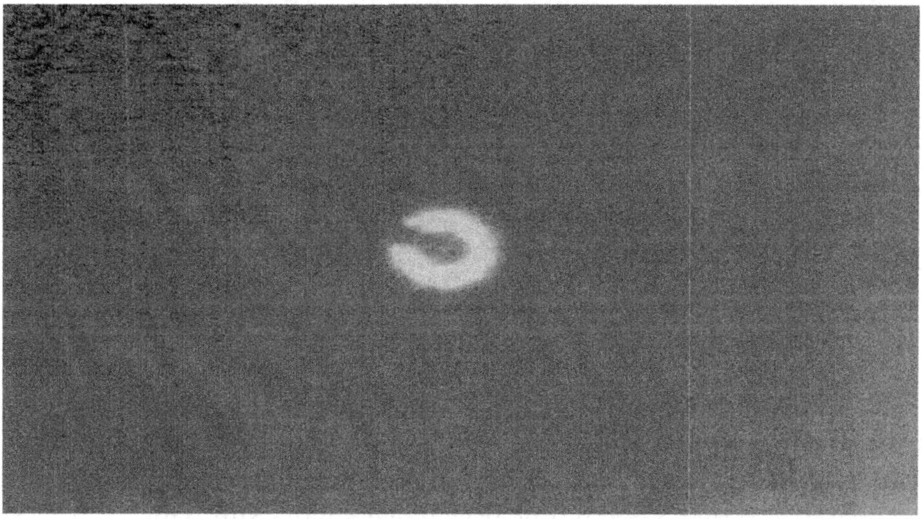

(Figure 9.1. Unidentified aerial object photographed prior to being shot down over the Yukon Territory, Canada, February 11, 2023. Image released by the Canadian Department of National Defence following joint NORAD operations; no debris was publicly recovered.)

Chapter Ten

3iATLAS AND THE BUGA SPHERE

The interstellar visitor known as 3I/ATLAS is now captivating astronomers worldwide as it hurtles toward its closest approach. Instead of behaving as a typical comet or asteroid would, 3I/ATLAS is expanding in apparent size and brightening at a pace that challenges established models. Its gas emissions are chemically unusual, and its light curve — the pattern of its brightness changes — is erratic and unpredictable. This points to a composition that appears different from anything native to our solar system.[114]

While many scientists cautiously attribute its anomalies to exotic ices or unusual fragmentation patterns, others are entertaining more radical ideas. Harvard astrophysicist Avi Loeb has raised the possibility that 3I/ATLAS could be an artificial object — perhaps a derelict probe, a fragment from an advanced civilization, or debris from alien engineering. If so, this would place it alongside 'Oumuamua and 2I/Borisov as the latest tantalizing clue that our cosmic neighborhood is occasionally visited by objects from other stars — and perhaps, other minds.[115]

3I/ATLAS was first detected by the Asteroid Terrestrial-impact Last Alert System (ATLAS) survey in early July 2025, flagged by its unusually high inbound velocity and an orbital path that could not be traced back to the Sun. This trajectory — hyperbolic and open-ended — marked it as the third confirmed interstellar object ever observed, after 1I/'Oumuamua in 2017 and 2I/Borisov in 2019.[116] Initial observations were puzzling: 3I/ATLAS appeared faint, yet its speed and inbound angle meant it had traveled for millions, perhaps billions, of years through interstellar space before entering our system.

As follow-up data came in, astronomers noticed that its rate of acceleration was higher than expected from simple gravitational influences. This was eerily reminiscent of the 'Oumuamua case, which also showed unexplained acceleration without visible jets of material. To Avi Loeb, this was more than coincidence. He pointed out that certain characteristics — such as the object's changing brightness and possible thin, reflective structure — could be consistent with a light sail or other propulsion mechanism not produced by nature.[117]

Avi and others have proposed the idea that this might be a technological object or an alien probe or part of one that has a hostile or spy functionality.[118] Some of the key reasoning behind this theory is the anomalies that are being observed. For example, the orbit plan alignment is argued to be an unlikely coincidence which could very well indicate deliberate targeting. Its trajectory allows it to pass "behind" the sun relative to the Earth, giving the possibility that it could covertly maneuver or deploy smaller probes while it is being hidden. Loeb suggests it could "release a series of 'mini probes' that arrive on Earth and appear as UAP or UFOs."

"It shows a glow ahead of the object, not behind it. If it were a standard comet...we would expect some dust to be shed off the surface...the dust would be pushed behind the object..." states Avi Loeb. Loeb also has postulated that the 1977 "Wow" signal that was received during August of that year many people to have been an alien signal submitted to the people of earth. He states that 3I/ATLAS is coming from within 9 degrees of the location of the original 1977 signal. Did it come from 3I/ATLAS almost fifty years ago?[119]

Between October 1st and 5th, NASA Mars Reconnaissance Orbiter and ESA's Mars orbiters Mars Express and ExoMars will observe this "comet" as it passes within 29 million kilometers from Mars.[120] The latest image is what resembles a giant tic tac UFO! Coincidence? On September 5th, the Defense Department was renamed to the Department of War.[121] It became the Department of Defense in 1947, the same year as the Roswell crash.[122] NASA, an independent civilian agency focused on exploration, quickly shifted its path and purpose. On September 25th, 2005 NASA quietly rebranded itself into an intelligence agency focusing on its new primary function, that being an intelligence gathering, counter-intelligence and national security agency. Why the immediate change?

Five days later, US War Secretary Pete Hegseth summoned approximately 800 top military generals and admirals from around the world to a meeting at Marine Corps Base Quantico in Virginia on September 30, 2025. There was a lot of speculation as to what they were meeting about. The video that was released may not have been what they were actually discussing. Hegseth and Trump spoke to these top leaders about their shape, fit an image? I am not buying it. The timing is extremely interesting.

On another interesting note prior to the rebranding and priority shift, July 11, 2025, NASA announced that it was going to be removing 400 social media accounts to a more streamlined focus going forward.[123] With still 300 accounts alive, NASA shut down accounts that include those dedicated to the Voyager mission and the Mars Curiosity and Perseverance rovers.

Back on July 9th in 2016 a bright lit white UFO like object can be descending towards Earth. The International Space Station abruptly ends its livestream when the object is abruptly observed. Another coincidence? On October 1st, 2025 we went into a Government shutdown.[124] Aside from critical missions and focus, NASA was officially shut down during this flyby. Weird timing I guess? Nick Pope former head of the UFO desk for the Ministry of Defence also chimes in on brand by stating

"It is not beyond the realms of science fiction that 3I/Atlas could be an alien spaceship of some kind...Of course, this could turn out to just be a comet...so interesting and incredible rare. The good thing is this is a testable hypothesis."[125]

Compared to previous interstellar visitors, 3I/ATLAS is estimated to be between 3.1 and 12 miles in diameter which is larger than Manhattan. Whether 3I/ATLAS is an icy wanderer or a messenger from another civilization, its brief passage through our skies will offer an unprecedented opportunity to study an unfiltered sample of the galaxy beyond the Sun's reach — and maybe, just maybe, a glimpse at technology not of this Earth. Could there be passengers aboard or is it something akin to our own space debris that is heading here? Will it be nothing?

If interstellar objects challenge assumptions about what enters our solar system, metallic spheres challenge assumptions about what may already be operating within it. Metallic spheres, orange orbs and orbs of living plasma have been showing up as early as the 1950s based on a report of one hovering over Fort Devens in 1957 along with orange orbs.

THE SPHERES

Could we be under surveillance? Patrick Jackson's book *The Sphere Network* explores the concept of an advanced, hidden system of intelligent orbs—categorized into three types—that operate globally as a protective network for Earth.[126] These orbs are theorized to be AI-driven drones exhibiting behaviors such as emitting electromagnetic pulses and influencing human perception, potentially explaining various UAP sightings and paranormal activities.[127]

The amazing discovery he made was by looking back at old photos of reported UFOs and UAPs and noticing small metallic spherical objects in a triangle formation in the photos alongside the photographed craft. The recent discovery of the Buga Sphere in Colombia, a metallic orb with mysterious engravings and internal structures, aligns with Jackson's descriptions of these spheres. The sphere's seamless design, symbolic markings, and reported behaviors mirror the characteristics attributed to the orbs in Jackson's theory, suggesting a possible real-world example of the phenomena he discusses.

Patrick Jackson's journey began with a surreal encounter in his twenties—while working in IT, he witnessed an apparition resembling a monk dissolve in a field. This sparked a shift from everyday skepticism toward investigating paranormal phenomena more rigorously. Combining his background in database development and reverse engineering, he began seeing patterns that many considered fringe or unexplained. I however do not. I believe as do many that he is onto something.

Jackson posits that many reports of poltergeist activity examples are creaking doors or electrical disturbances; these aren't supernatural in the spirit sense, but rather anomalies triggered by technologically advanced spheres. His theory suggests:

- **Type 3 Spheres:** Small units found in haunted places, generating localized effects or distractions.

- **Type 2 Spheres:** Mid-sized beams hovering in skies, relaying signals upward.

- **Type 1 Spheres:** Large, fast-moving crafts often seen in V-shaped formations during UFO sightings.

Collectively, these form a cascading communication or surveillance system—akin to relay satellites—able to triangulate and monitor targets with precision within seconds. In The Sphere Network (published February 15, 2025), Jackson expands his thesis into a full framework for a hidden, AI-powered planetary defense system composed of these spheres. The book draws on eyewitness reports, technical illustrations, and his investigative methodology to suggest that these orbs have been operating—covertly—throughout human history.

Patrick Jackson turned his tech-savvy background and a strange personal experience into a structured hypothesis: luminous spheres—ranging from small to large—are an engineered network, not paranormal ghosts. These spheres might relay data, monitor, and even protect Earth via stealthy triangulation. While unconventional, his theory has gathered attention both online and from curious academic voices.

Wouldn't it support Patrick's theory if one of these were to fall from the sky as if it were a gift? In March 2025, an intriguing incident occurred in Buga, Columbia. A metallic orb which has now been dubbed the Buga Sphere was observed hovering and maneuvering in a strange zig-zag trajectory that defied known flight dynamics. It is reported that it got very close to some power lines and was knocked out of the sky falling to the ground. A farmer with his bare hands retrieved the metallic soccer ball-sized orb. Mexico's National Autonomous University (UNAM) began studying the object.

While performing microscopic scans, the researchers found the sphere contains a labyrinth of fiber-optic-like cabling, a central "chip" was sandwiched by 18 microspheres with three layers which suggests the potential for signal processing capabilities. Similar to UAP and UFO reported landings or encounters, there appears to be a physical trace that was left behind. On the ground, the orb left behind a peculiar aftermath: a patch of soil and grass that completely died off and showed no signs of recovery. I immediately think about the farmer that held this object in his hands and what physical effects may occur with his physiology over time.

There was no radiation detected so researchers are now leaning on the theory that maybe the object is accessing an unknown energy field that may have drained the area's moisture. What is mystifying is that the object also indicates that it is increasing in its physical weight! The sphere's weight appears to have increased from 16 to 20 and then to 22 pounds which has sent the scientists into a flurry as this has now prompted theories that the sphere is using a futuristic gravitational force or forces to lighten the load while in flight!

A video has emerged showing an individual chanting Sanskrit mantras and Vedic hymns near the sphere, and claim the orb vibrates or responds in rhythm. While skeptical scientists like Dr. Julia Mossbridge dismissed the sphere and its unique writing as being something reminiscent of an art installation or an object that is man-made.1[29] Immediately we now enter the realm of science fiction with a theoretical paper by P. Morcillo titled "A Unified Framework for the Buga Sphere: Quantitative Validation of a Negative-Mass Model Governed by Topo-Temporal Physics" proposes which could completely shift our concept of motion, energy and time and how we might be able to manipulate it:

- The sphere exhibits a negative-mass effect, drastically reducing its perceived inertia.

- It may move without visible propulsion and produce a stable cooling effect (-100 watts).

- This phenomenon may arise from interactions with a fractal-like structure of space-time, something beyond current physical understanding.[130]

Now, there are some strange etchings, glyphis, symbols and carvings on the outside of the Buga Sphere. No one has been able to understand what the symbols represent. I employed ChatGPT, uploaded the widely circulated photo of the sphere and this is what it spit out as an interpretation of what the symbols and shapes might mean. In the center is a microchip-looking symbol. It is astounding.

Central Microchip-Like Symbol

Translation: “Source Core”

Symbolizes a unifying intelligence, possibly a consciousness hub or data memory. It may represent the origin of information or the “heart” of a system, akin to an artificial or cosmic intelligence.

Circular Glyph Ring

The glyphs around the center are not from any known human script, but based on symmetry and repetition, we can interpret them symbolically.

Possible Translation:

“Encoded in light. We are the Watchers of Balance. Protectors of Living Code.”

• “Encoded in light” refers to knowledge or consciousness embedded in frequencies or energy.

• “Watchers of Balance” could indicate a custodial role in maintaining harmony across

systems—biological, cosmic, or energetic.

• “Living Code” is a term often associated with DNA or the fundamental essence of life.

Outer Star Shaped Pattern with Humanoid Figures

These resemble stylized humanoids surrounding the sphere in unity or motion.

Interpretation: Represents collective consciousness or a council-like unity among beings. It implies cooperation, equality, and a shared orbit around a common truth or intelligence.

Triangle-with-Circle Symbols (in alternating pattern)

Often interpreted as “portal” or “observer” symbols in esoteric language.

Meaning: They may suggest ***interdimensional*** travel, awareness nodes, or monitoring outpost

Complete Symbolic Translation (Poetic Form):

We are the bearers of encoded light.

From the central source, all life flows.

We orbit the truth in unity and purpose.

Guardians of the balance, observers of time.

The code lives. Protect it well.

There have been recoveries of these metallic orbs in the past as well. Did they have the same glyphs or designs?

1. The Betz Sphere (Florida, 1974) – A family discovered a metallic ball that exhibited strange behaviors, examples include moving on its own, returning after being pushed, and emitting vibrations. It drew military and scientific attention but was never conclusively explained.[128]

2. Klerksdorp Spheres (South Africa) – Found in Precambrian rock, these smooth, grooved metallic objects were claimed by some to be out of place artifacts, though geologists largely consider them natural formations.

3. Vietnam War “Foo Fighters” – U.S. troops reported glowing orbs and metallic spheres in the sky, often assumed to be advanced surveillance craft, though no definitive source was confirmed.

I had the privilege to appear on the streaming platform GAIA with George Noory and one of my friends Josh Golembeske is in charge of production. He recently had the opportunity to attend the press conference in Mexico regarding the Buga Sphere and was one of the few to actually hold it in his hands! He was gracious enough to allow me to interview him for the book and share his thoughts on the Buga Sphere and other recent discoveries.

Josh stated that if this Buga Sphere turns out to be a real UAP which he said he is learning towards after his time in Mexico for the press conference, then that moment where he is holding it in his hands is a life changing event for him and essentially all of humanity.

Josh jumped right into what his gut was telling him. "I headed down there skeptical expecting this object to be some sort of a hoax, but after interviewing the witnesses and the engineers and seeing all the science that had been done on it and what they were aiming towards...I realized oh my God, one this is not a hoax and this is a real object that was really in the sky. Number two - what really is it? Why was it in Buga, Columbia? "

"Right now if I had to guess, if I had to put money on it...I would say reverse engineered high tech craft that was doing surveillance from another competitive country in South America. It does however fall in line with other spheres that have been seen around the world that indicate that they are connected to NHI or Non-Human Intelligence."

He continued, "the science behind it is interesting and a little hard to explain, anomalous would be the term used on how this sphere actually operates." Is it true that the sphere started to add weight? "Yes, over the course of several months it increased from 2 kilograms to 10. One of the engineers theorized that it operates using a magnetic field as its propulsion system and it was slowly gaining back weight as it was losing access to this field. The scientists and Jamie Mussan and his team are now trying to determine how old the sphere is and this type of measurement is very costly when you are dealing with something that is non-carbon based."

What was the press conference like? I saw that Dr. Steven Greer was present. "It was like something out of a movie, Greer was there with a bunch of folks from NASA, scientists and contractors that wouldn't go on camera but talked to us on the side.[131] They were there while all of this analysis was being done. The National Autonomous University of Mexico was studying the object's strange properties along with the engineers."

"U.S. Congressman Eric Burlison, a Republican from Missouri was there taking it very seriously. Attorney and UFO disclosure supporter Daniel Sheehan was present as well. It was the who's who of UFO researchers and there was an incident where two people tried to take the sphere at the end of the conference about a day later! They were dressed up posing as security guards but they were stopped before they could take it away from the real security team."

Who do you think they were? Josh said they were there about five minutes before it happened. "We talked about it afterwards and we all agreed that if they were from one of these UFO legacy program 'clean up teams' trying to keep it a secret, they wouldn't have been caught or stopped. They would have been successful. We don't know who *they* were. They disappeared."

Josh goes on to say, "I believe that these spheres are some kind of extraterrestrial network for surveillance that could be interdimensional in nature, because they are seen all over the skies and they blink out and disappear and reappear on radar. I don't believe what created them is though, I think the lights, the orbs are different from the spheres and yes those are 100% interdimensional and connected to religion, angels, demons whatever you want to call it."

As for Buga, Colombia, it isn't widely known internationally as a UFO hotspot as are parts of Brazil or Argentina. However, Colombia has had notable sightings, especially in the Andes and rural zones. Nearby areas have reported orbs and anomalous aerial phenomena over the years, especially near military zones or energy hotspots. But right next door - there is activity that has been going on for a very long time, hundreds of years, maybe over a thousand.

I recently met with a couple, about a month ago, who came to one of my talks in Warwick, Rhode Island on the topic of New England UFOs. They were in town from Guatemala to visit his ailing father. He had reached out to me via X about showing me videos and photos that he had taken personally above his AirBNB property in the remote jungle. I was a bit skeptical but once I had a chance to meet him and discovered that he was a former monk and school teacher of thirty years and had moved to his wife's homeland for a different way of life and that they were filming dozens of these UFOS and UAPs hovering over his property on a nightly basis - i was in!

I saw the most compelling UFO videos I have ever seen. One of them was a craft that seemed egg-shaped and horizontal and emitted a sound that would be the equivalent of a digital bird. It had a very digital and computerized sound to it. There would be a dozen of them showing up with a small rogue orange orb that most often would be left behind and one of the craft would come back to urge it to follow. The couple, who I will keep anonymous for now, has invited me to come down to the jungle to witness these for myself.

One of the most incredible findings is that they reached out to Nick Pope who gave them a phone number. They talked with someone from DC or Virginia that was associated with a UFO desk. Their property is in the remote mountains and they have about twenty tribes that live close by and do not read or write. They were persuaded to take a picture that included all the different sizes and shapes of the UAP and UFOs that have been reported over the years.

The style or type of UFO that they have been seeing for over a thousand years just happens to be... the tic tac UFO! Ross Coulhart has recently come forward stating that he has been told by reliable sources that what Navy fighter pilots were interacting with was not alien tech by US technology, specifically Lockheed Martin.[132]

This reveal to me which happened within a week of Ross' questionable statement led many people to believe he had been prompted to push listeners and viewers into yet another cover-up. Maybe it is based on the alien tech design, reverse-engineered...but it might also be the real thing based on its maneuvers and trained pilots' reports of its capabilities. To be honest, although excited that I was invited I am a little nervous to make a trek out there and not come back if you know what I mean.

Chapter Eleven

INTERDIMENSIONAL BEINGS INTERACTING

For whatever reason my experiences continue to increase across all these various phenomena. As with most people, it starts with a fleeting glimpse or a distant sighting, finding tracks or signs that indicate there is more going on but then there are interactions that are undeniable, they escalate and become paradigm shifting to our belief systems.

Moments when something that is supposed to be a myth, a fantasy or folklore out of nowhere becomes real, responsive and capable of engaging with us, the observer. In the following cases we are seeing an enforcement of protecting their territories or boundaries that we cross because these non-human intelligences operate within overlapping environments that intersect with our properties and homes we live in and when we enter their domain they let us know.

While these experiences are personal and subjective they have a consistency across independent witnesses that mirror patterns long documented in both folklore and modern-day high strangeness research.

In Wampanoag and other Algonquian traditions, Pukwudgies are often portrayed as once-helpful spirits who turned against humans after feeling disrespected or ignored. One story says they originally lived alongside humans peacefully, but grew jealous of the affection shown to Maushop, a giant culture hero of the Wampanoag. When the people began ignoring the Pukwudgies in favor of Maushop, the creatures turned mischievous and later deadly.

The word Pukwudgie comes from the Wampanoag language and roughly translates to "*little wild man of the woods that vanishes*." These creatures are part of Algonquian folklore and are described as small, beings resembling a troll or goblin that has magical powers.[133] They are known for their trickery and sometimes malicious behavior of luring unsuspected victims to a cliff's edge with the intention of leading these innocents to their death. Whereas other legends state that they are guardians of nature. The Ojibwe people believe that the creatures are friendly and play harmless pranks.[134]

They're seen as deeply tied to the land and forests of New England, particularly areas such as the Freetown-Fall River State Forest in Massachusetts, Monsterland and the areas surrounding Leominster State Forest in central Massachusetts where many modern encounters are still reported. For over the past several hundred years there have been sightings of them from the East Coast to the Great Lakes.

The Pukwudgies stand about two to three feet tall, have gray skin and large ears with the ability to shapeshift into different creatures. One of these looks resembles a porcupine from the back and a half-troll, half-human from the front. There are legends that speak of their ability to turn invisible, make people forget things, steal food, knock travelers off paths or shoot victims with magic arrows.[135] [136]

The most recognized modern-day story emerges from the Bridgewater Triangle in Massachusetts in 1990 where local resident Bull Russo was walking his dog late at night when he allegedly spotted a three foot tall humanoid creature that was covered in black hair and spoke in a bizarre language, trying to lure him into a nearby swamp. The area is known for unexplained deaths and disappearances. The local activity arguably originates out of the Hockomock Swamp, which means "*Where the spirits dwe*ll".[137]

I was surprised to have someone reach out to me because one of their family members was witnessing something he didn't understand when he was out bow hunting. The exact location hasn't been revealed to me as of yet, but I have a good idea of where it might be and it's right here in Leominster. This hunter is unwilling to speak with me face to face but his brother has been the bridge between the two of us. The hunter would walk to this location near his house and began to notice what looked to be a small being, peering out from behind the trees and shrubs watching him as he laid in wait to hunt. After several visits to this location...the strangeness began.

Whenever he returned to this hunting area it wouldn't take long for this being to show up. It would without warning appear, but using the flora, trees and shrubs to camouflage itself. It would then vanish and reappear on the opposite side. He thought he was beginning to lose his mind. How could this thing disappear and reappear in another location in less than a second? He was *encouraged* to take photos of this being if he could. This photographic evidence would quickly extinguish the paranoia that this was simply a concoction or creation of his mind. You can't photograph something that isn't there in the first place. "The hunter was finally able to capture the being in motion as it fled the area (see Fig.11.1)"

With each hunt going forward, this being, this creature, would get closer and closer to him. He was finally able to snap a photo of the being, likely aware of his intent as he captured it in motion as it ran away. I lost my breath when I saw this thing. It was an elf or goblin-like entity. Large nose and large solid black eyes. It looks as if it is wearing a hat and maybe even boots?! I believe that this hunter was interacting with a Pukwudgie! Others may also see something peeking from behind the tree which bears the features of the head of a Bigfoot or Sasquatch.

One day this hunter was out there and this Pukwudgie showed itself. Materialized was the term that was used, but it was instantly accompanied by a native american! They began to have a conversation in front of the hunter! Both were clearly visible. The pukwudgie began to talk in a strange language, turned and looked at the hunter and was pointing directly at the hunter. The hunter could hear it talking in a strange language which he believed was of local tribal origin. He got the sense that he was in this being's hunting area and it was complaining to someone who might be in charge. Sounds very much in line with the legend described at the beginning of the chapter doesn't it?

The hunter described this and you can't help wondering if it was as if he was looking at or into another timeline as he mentioned the native american seemed to be from another timeline. But how are they interacting with him now? Is this a portal area or location where timelines intersect. The reason I mention this is because on another hunt he described seeing a "dinosaur' moving through the woods and using the term something out of "Jurassic Park", Again almost as if he was seeing or allowed to see some sort of timeshift or other time period.[138]

This was all too much for him. Well, he decided to stop hunting for good when he made his way back from the hunting area walking back home. When he turned around he noticed that this Pukwudgie had followed him home and was watching him from the backyard. All the images, the emails that he sent to his family members who contacted me were deleted out of fear. He mentioned his worry of having Men in Black or some government agency would be knocking on his door. Luckily, I have one of the images. It may be the only surviving one. This was taken in late summer, August 2023. I have also zoomed in on the subject in another photo below it. What do you see?

On September 12, 2023 my neighbor came knocking frantically on my door at 7:30 am in the morning. The previous day, 9/11 we had a strange rain storm that dumped a foot of rain in four to five hours causing massive flooding of our brooks that run through the backyards of a lot of houses including mine. The damage was so significant that FEMA, the Federal Emergency Management Agency, would be involved. Within a few hours of the rainfall, the private road that connects my house to my neighbors behind me, which abuts a state park, had collapsed. The brook that runs underneath had flooded and the rapids took it down.

My neighbor was walking around and surveying the damage. As he looked down for a moment, from his peripheral to the right, he witnessed something bipedal, upright, covered in black hair and about three to four feet tall running from one tree to another. It was using the trees for cover. It sent a shiver down his spine. My neighbors first thought was, “I need to go tell Ronny! This is right up his alley”.

His unexpected visit quickly turned into an interview. He did mention, despite not seeing anyone around, sensed he wasn't alone and had the awareness that he was being watched. This sighting was confirmation of that suspicion and for it to happen so quickly and so close to my house within a few minutes, it was basically served to me on a silver platter.

I walked out to the area that he stated he saw something running between the trees. As I started to explore a little bit just in the immediate area I noticed small footprints resembling that of a child, fresh, in the mud alongside the brook, right where the road had fallen into the brook. It had just rained and these looked to be created within the last hour or less. Trolls and a bridge. Weird. Our new house and the area is living up to its rumors. The Barrett Park area was restricted due to the flooding and was subsequently taped and marked off.

At 8 or 830 in the morning, myself, Amy and the kids went to explore more of the damaged area. Pieces of the asphalt road were found yards away as we followed the brook's newly created path. Elements of the road and rocks we had never seen before, exposed from the erosion, were being swept away by the raging rain.

The brook quickly metamorphosed into a mighty river. The dam at Barrett park was in jeopardy of collapsing so several homes were evacuated and no one was allowed in the park. We seemed to be the only humans present in the woods as we made our way down further until we approached the pond.

As we approached the water, the trails were soaked and in some spots submerged underwater. We got closer and we all noticed footprints. Later that night my daughter's boyfriend filmed a UFO hovering in the sky over these exact woods. Her bedroom window faces the park!

The pukwudgie that the hunter witnessed looks as if it has boots but these prints suggest a smaller sasquatch rather than a pukwudgie or was it something else? "Fresh impressions along Colburn Pond in Barrett State Park, followed by a trackway down the brook, followed later that night by an anomalous set of lights hovering and moving over the same woods (see Fig 11.2)."

Creature from Another Realm?

October 2019 in Salt Fork State Park, Ohio, myself, Matty Blake from the History Channel and WAAF Boston Rock radio, Jeff Byers (Owner of Creatureplica) and Dr. Rebecca Foster, a clairvoyant medium, attended the Ohio Bigfoot Conference. The park is well known for its encounters and is considered a hotspot reminiscent of Monsterland.[139] We had the opportunity to interview the legend Bob Gimlin from the Patterson/Gimlin film while we were there for my short-lived Monsterland Podcast. It was his birthday weekend and what we all experienced together later that evening was simply mind-blowing. It was around 1AM when we were leaving the convention center and walking towards the cabins where we were staying, which was a relatively short walk.

There was a large parking lot and a forest between the two. As we were walking back to the cabins, I noticed in the woods what looked like a small orange orb moving back and forth at about our eye level, so it was quickly noticed. Dr. Foster started walking in that direction and we followed. This orb or reddish-orange light, was from our distance and in my estimation smaller than normal. I would typically see these the size of basketballs. But this one, maybe the size of a softball or baseball? It was hard to tell so we moved closer. This light was moving back and forth in such a way that I was certain it was trying to get our attention. “I think we got something,” I called out. I pointed in the direction of the orb dancing in the dark woods yards away from our position. We got closer for a better look.

The four of us walked up to the edge of the woodline and watched as this orb danced around in front of us but stayed within a five foot radius. I began to call out for the prankster to reveal themselves. There was no response. I picked up a small rock and threw it into the direction of this strange light hoping for a response. We could hear the rock land onto the leaves and branches below the light and we could hear it roll a few feet. It never connected with its intended target. No response.

As we watched, it appeared that the orb was shifting and splitting. It then became two orange-reddish eyes about seven to eight feet off of the ground. They looked as if they were moving back and forth behind the tree!

My initial thought, which Matty then shares with the group that same thought, is that this might be someone with a red lamp on their head maybe trying to mess with us. We were at the Ohio Bigfoot conference of all things. But two red lights now in order to generate fake eyes? It didn’t make sense. I began to get the chills. Dr. Foster had an idea and began to play a recording through her phone’s speaker. The recording was of a native american shaman saying a prayer in the Cherokee language.

I could feel the energy from whatever this was emanating out from the forest and towards us. I then invoked this thing to show itself. "Come on, show us what you are" I called out into the darkness. I had my hands outstretched and almost in a squatting position, I motioned with my hands for it to come forward and show itself. (I tend to do this a lot apparently)

And to our surprise these two glowing eyes formed in front of the shape of a dark eight foot shadow.[140] I could hear Jeff excitedly tell all of us to shut off our lights. "Go dark," said Jeff. "Shut off your lights." We all clicked off our lights and stood there in the blackness unaware of what we were up against. He then said something that freaked me out. "Get ready, they are going to start surrounding us." he then stated calmly.

"They!? I responded. And sure enough, we began to hear the sound of branches breaking and leaves rustling off to our right. The shadow figure in the shape of a tall man kept Dr. Foster looking straight ahead while it rhythmically and quite mesmerizingly danced in place from right to left, back and forth almost hypnotically.

I realized after that this was some sort of technique to keep our attention from the others who were quickly flanking us on each side. I then heard the sound of footfall and something walked to the left of us. When I looked over in that direction there was a small stream of moonlight that was cascading down onto the forest floor and I watched something bipedal and with grayish hair, barely a neck, huge shoulder and estimating seven to eight feet tall, taller than a normal human being. It was awesome. It was covered in greyish hair as it walked into that light. I didn't see it before or after that stream of moonlight almost as if it wasn't registering with my eyesight or it was so dark I just couldn't see it.

When my gaze went back to where the glowing eyes were situated and where Dr. Foster was kneeling down and watching this thing rocking back and forth, I noticed that I couldn't see the eyes anymore or the shadow. I quickly aligned myself a few steps to the right side of her so that I would have her similar viewpoint. I scanned but I didn't see the eyes or shadow anymore...but she was looking straight ahead with her eyes fixed forward.

"Where did it go?!" I asked excitedly. She replied calmly while still looking straight ahead "It is still there. Get ready. It is going to move from right to left of our vision in about fifteen seconds. It is going to run in front of us." I looked at her in disbelief trying to understand how she knew this. "How the hell do you know that?"

Dr. Foster then looked at me and said to me straight-faced. "It just told me."[141] And I immediately knew that it had communicated to her telepathically via her mind. And as if counting down to a rocket launch, 10, 9, 8... we waited and at the fifteen second mark, this eight-foot shadow, which we couldn't see anymore revealed itself again to us as it ran from right to left in front of us and covered what seemed close to twenty yards in seconds.[142]

We could see the silhouette of its form against the moonlit forest and silhouetted trees behind it. Here is the weird part, we didn't hear a sound when it "ran"! It moved as if it were a ninja through the forest and the fact that it didn't generate footfall or any type of sound was alarming and instantly created a sense that we weren't dealing with an unclassified primate. And then...nothing. The show was over. I was in awe. Did a bigfoot somehow float across at neck breaking speed and silent? Was it something else entirely? Well, I saw more than one and we were pretty confident there were three of them. But here's the bigger question, did it seriously just communicate with Dr. Foster telepathically?

It was another event and personal experience that cemented the fact that what the world knows as Sasquatch is quite possibly something entirely different. More people are having these strange or paranormalesque encounters with these beings.

After this energizing and awakening encounter, we all walked back in disbelief and somewhat shockingly reviewed and analyzed between one another what we just had experienced. I believe Jeff mentioned that the show "Monsterquest" had filmed here and near these cabins years ago and Finding Bigfoot as well due to its high activity and sightings. As we got closer to the cabins there was a trail off to the left on the opposite side of the forest from where we were and directly situated behind the cabins. We thought it was over. It wasn't.

The trail went straight up towards a razor-wired, fenced-in electrical power unit that was giving off a very low sounding hum that seemed deafening, especially this late at night. As we walked past, Jeff stopped and then started to walk up the trail. I looked back at him in disbelief. Why is he going up there? I thought to myself.

That sound is so loud, you aren't going to see anything. We were not too far from the cabins. Is he going to relieve himself I thought? We watched from the bottom of the trail and we started to walk up a little bit closer as Jeff seemed to make it to the top, he then surprised us all with a quick move into the forest to the left of the trail and swiftly disappeared from view. We looked at each other. We can hear him walking onto the forest floor on top of the hum, leaves crumbling, branches breaking as he moves deeper inside the woods.

After what seemed to be only eight to ten seconds we all heard what sounded like a lot of commotion and thrashing coming from his direction. We heard running and we looked as he launched himself out of the woods and onto the trail and was running back down towards us like a track star! There was a wave of laughter or hysterics that seemed to penetrate the three of us down below like a wave of energy. It sent a quick message.

RUN! Immediately all three of us took off with Jeff running behind us heading our way. We were instantaneously projected into panic and we were thrown into a state of fright or flight. We started running away and towards the entrance of the trail.

As I looked back I saw what Jeff was running from. It was HUGE. I could see the silhouette of something very large in stature and height bent down on all fours but it looked unnatural in the sense that it wasn't a four-legged animal but something tall, large, bipedal, like a sasquatch, trying to look like something else. It was too thin to be a black bear and the arms were extremely long. I had the sense I was looking at a Bigfoot..again. It was blacker than black almost shadow-like where we couldn't make out any features. We just experienced what sounded and looked like three of them on the other side. Had one of them simply followed us or was this a fourth? We safely scurried to the cabins and stayed up for a little while longer until we settled down. About an hour later we all headed for bed.

The following morning we gathered together for breakfast and recalled what transpired just a few hours earlier. Dr. Foster shared with the team her dream was more real and lifelike. She was looking through the eyes of something tall and hairy as it was traversing the forest. It was now approaching a lonely, single cabin in the distance. She was looking through the eyes of what she interpreted as a sasquatch as it stealthily approached the glow of one of the exposed cabin windows.

While Dr. Foster was seeing through the eyes of this being, it approached the window and it knelt down so that it could look inside. To her surprise, she could see *herself* sleeping on the bed! She jokingly remarked to all of us that she had this "aha moment" during the dream when she saw herself laying there and she was somehow looking at herself. Her response was simply, "Great, they can do that shit too!?" We all laughed.

Matty and I had to leave later that morning. A thirteen hour drive from Ohio back to Massachusetts but with a few supernatural stories to share back home. Later that night, Dr. Foster and Jeff decided to venture up that trail where Jeff had been bluff charged. After hearing about what happened next, I wished we had stayed one more day to experience it.

Dr. Foster told us what transpired. “Jeff and I went back to the trail and we didn't have anything with us. No flashlights, no equipment. It was super quiet and then it shifted at once to being very creepy as we stood there in the darkness at the bottom of this trail. We went completely technology free.” It wasn’t intentional. She then goes on to say that she had this thought. “Then I was thinking, maybe we should just get down and look like we are enjoying ourselves.” She shares this idea with Jeff. “Okay.” he said following her lead.

“So we got down to the ground and she wanted to hint to whatever was watching them that they wanted to throw something. So we started collecting rocks on the ground near us and started to toss them back and forth at one another.” Dr. Foster then looks up and sees the light refracting off of this thing that was not there a second ago. “It was huge and black!” she said. “Jeff, do you see that thing that is up the trail right there?” He looks and says, “Yes, you mean off to the left? Yeah.” So they are now aware of something very large watching them play with the rocks.

They didn’t hear it approach them in their area. It just seemed to appear out of thin air. They tossed the rocks back and forth between each other for a few more minutes and when she looked towards this thing it began to move closer to them but it wasn’t moving...it seemed to almost imitate floating towards them.[143] “Jeff, is that *thing* getting closer to us?” And he said, “Yes, it is getting bigger.” Dr. Foster said she didn’t want to stick around anymore.

She estimates that it had to be at least eight to ten feet tall. They could see the outline of this dark mass under the moonlight. "I have full faith in everything that she is feeling", said Jeff. They realized that there was no one else in the park nearby. Jeff said that they both were drawn to go back to this same spot again earlier in the day and now here they were and they were not disappointed.

Matty recalled about our experience together and stated, "I know now that after leaving Ohio, I didn't have an experience with something biological. It seemed to be more paranormal. Capabilities that seem to not be of this earth". Dr. Foster chimed in and said, "I think along the same line as you but I don't want to say 'paranormal' and exclude the aspects that this is something that is humanoid and animal, so I don't want to use that term. However it presumably falls along the lines of Native American culture and those stories. What we experienced has some form of spiritual properties to it because those things are so quick and silent."

"They are going through something, they are traveling here somehow." she said. "Something interdimensional?", I ask. Dr. Foster says, "Yes. and the one that we saw in Ohio has asked me about ten times since then...when are we coming back?" We all freaked out. "What?! So you are experiencing what is called 'mind-speak' that others talk about after having a sighting or an encounter with a sasquatch?" She responded emphatically "Yes. I think that they are historical, humanoid with the ability to travel through dimensions or portals of some kind. There is no excuse for a 400 or 500 pound person or animal to traverse dry leaves and not make a sound."

I haven't made it back to those woods since that epic experience, but I think a return trip to South Fork State Park in Ohio is overdue. Zooming out and looking at these experiences for example the Pukwudgie encounters in New England or the Sasquatch interactions in Ohio they may look as if they are separate occurrences but they don't read as separate phenomena. They are variations of the same underlying intelligence expressed through culture, geography, and perception.

These experiences and encounters have shared traits and some examples would be immediate silence, instant manifestations, telepathic as well as audible communication. These beings have an ability to manipulate attention and the space around them by going invisible, disappearing and reappearing or forming out of a ball of light and even interacting with us in our dreams before or after an encounter. If the beings are interdimensional in nature, then folklore, modern eyewitness accounts and personal experiences do not contradict one another but highlight fragments of an older and continuous relationship with humanity.[144]

(Figure 11.1a Eyewitness photograph of a small humanoid entity partially concealed by foliage, Leominster, Massachusetts, August 2023. Identity of the witness withheld per request.)

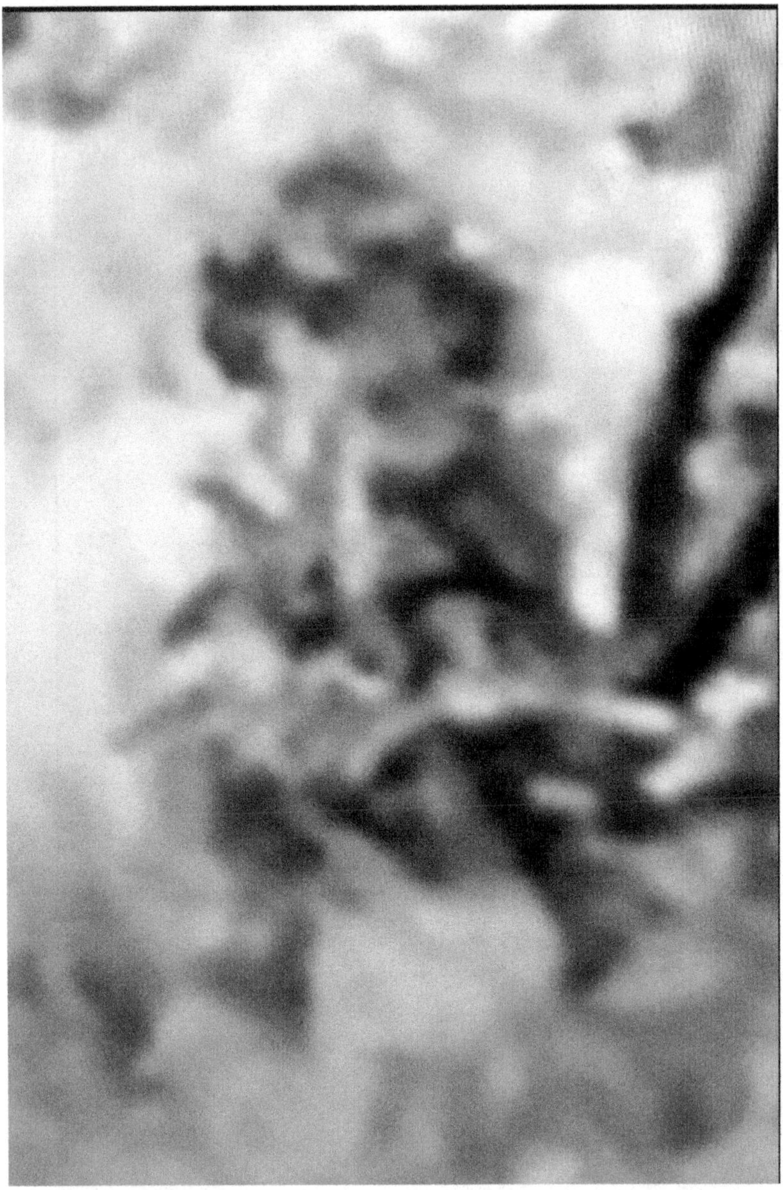

(Figure 11.1b Close up, cropped and enlarged detail of Figure 11.1a highlighting facial features and apparent headwear; no digital alteration beyond contrast and scale adjustment)

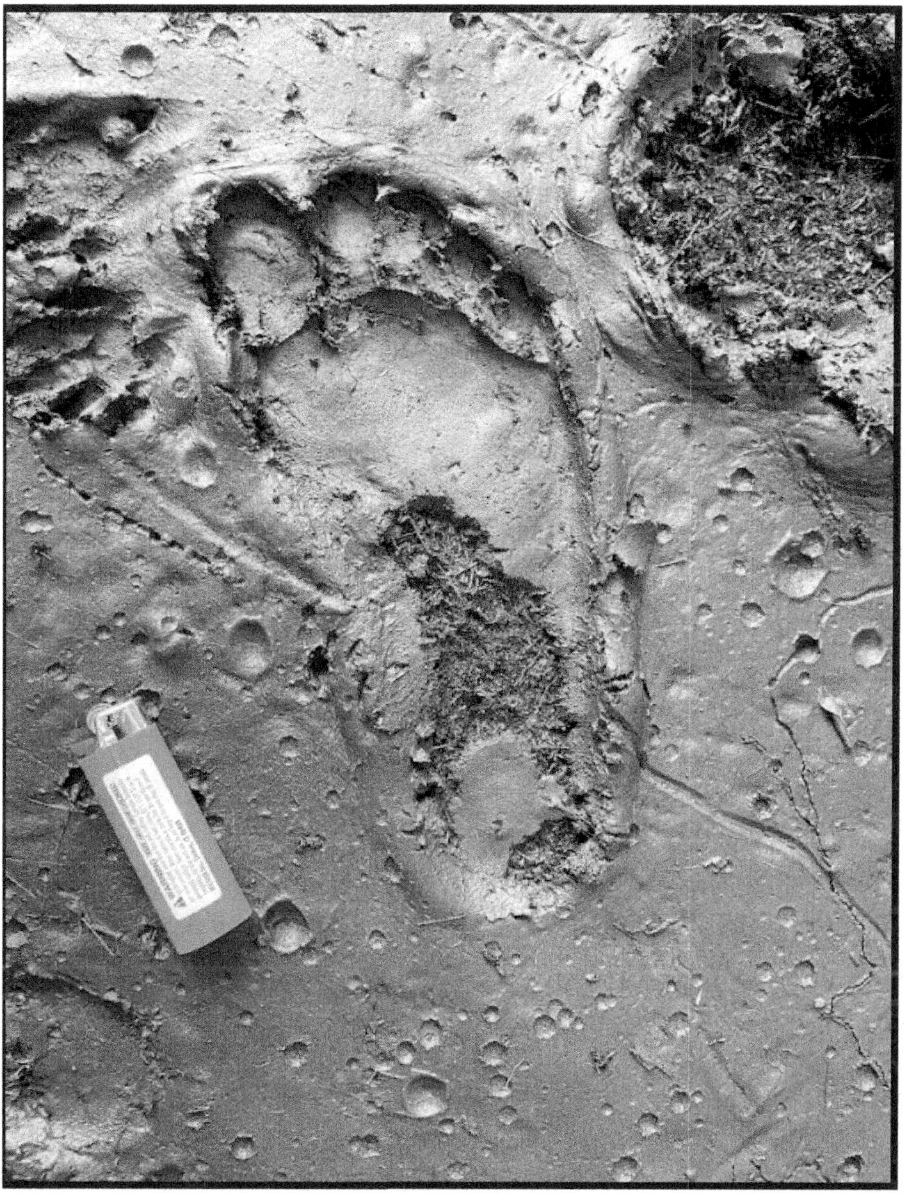

(Figure 11.2 Footprint and anomalous light (no photo) in Barrett State Park, Leominster, Massachusetts, September 12, 2023. Photograph by the author)

Chapter Twelve

UFOS AND THE MILITARY

Lue Elizondo had stated unequivocally that the Pentagon's UFO investigators were very interested in the after effects of their military personnel when they encountered a UAP or UFO. The body chemistry of the individual changes and specifically a region of their brain often associated with precognition and advanced sensory capabilities grows![145]

During an interview with News Nation Lue doesn't think that whatever is behind the UAP phenomena has the best intentions for humanity. "There are enough reports that substantiate that not all these interactions are necessarily benign or peaceful," he said. "Some people leave terrified, some people leave injured. There are U.S. service people who are on 100% disability right now from the U.S. government because of an interaction with UAP. And it's in writing, by the way, from the U.S. government because of their involvement in an incident."

An example would by The Veterans Administration (VA) has granted full medical disability benefits to Airman John Burroughs for injuries to his heart and eyes that he claims to have suffered during the famous 1980 UAP incident in England's Rendlesham Forest, a series of reported sightings of unexplained lights just outside Royal Air Force Woodbridge, which was used at the time by the United States Air Force.[146]

Some of these negative after effects to pilots, special forces operators and civilians are mysterious brain seizures. This was intriguing to me because last October I suffered two major seizures and spent several days in a Boston hospital. I still don't know what happened and I am due for a follow-up soon. Is it just a coincidence that remote viewers and those associated with UFOs, studying UAPs are experiencing this? Scott Davis, the New Jersey police officer when he tried to remote view the drones in New Jersey and was up against a mental and physical wall that he couldn't penetrate. Is it some sort of defense mechanism?[147] [148]

Just to be clear, there are alternative explanations that have been considered and they are still being reviewed, such as advanced foreign surveillance or weapons systems being deployed, directed microwave or radio frequencies impacting individuals, before attributing these neurological injuries solely to non-human intelligence or interdimensional interaction. However, we are seeing patterns again when we talk about infrasound potentially being emitted from bigfoot encounters, have extreme headaches that are only quelled when the person asks for it to stop and apologizes or blacking out when in close proximity to an orb or paranormal experience. We are seeing a common thread with the mind and consciousness being affected, altered or manipulated.[149]

These are the same patterns that continually persist across generations and locations known to have a history of unexplained phenomena and those that have encountered it can be affected by it no matter where they are. The conventional explanations do not fully account for the highly specific brain injuries that are observed, particularly the damage to the caudate-putamen region in otherwise high-performing individuals, nor do they explain why similar symptoms appear across what many people or unrelated or not connected like UAP encounters, Havana Syndrome, remote viewing programs and or historical military and private contractor experiments involving electromagnetic fields.

David Grusch, the former U.S. intelligence officer who became a prominent UFO whistleblower, has spoken publicly about suffering from serious head trauma. In interviews, including with NewsNation in 2023, Grusch mentioned that he developed post-traumatic stress disorder (PTSD) after a traumatic brain injury (TBI) he sustained in Afghanistan while serving in the Air Force. He was an intelligence officer attached to special operations, which put him in high-stress, high-risk environments.

He didn't go into a lot of medical detail, but he implied that the head injury and PTSD were tied to events during his military service, not related to his UFO work. Some media tried to use his medical history to question his credibility, but supporters argue that many veterans have similar injuries and are fully reliable.

Yes, David Grusch has publicly said he suffered seizures after his traumatic brain injury. In interviews, he mentioned that after his TBI, he dealt with post-concussive syndrome and seizures, which led to ongoing neurological issues. David Grusch said his seizures began after he suffered a traumatic brain injury (TBI) while deployed with the U.S. Air Force in Afghanistan. After the TBI, he developed post-concussive syndrome, seizure disorders, and PTSD.[150]

This was part of why he was hospitalized for mental health treatment at one point, which some outlets tried to use against him after he came forward about UFO whistleblowing. He's been open that the injury and related health struggles were from military service, not from anything involving his UFO disclosures.

Here's a simple timeline for David Grusch's head injury, health issues, and whistleblowing path:

- 2010s (exact year unclear): While serving in the U.S. Air Force in Afghanistan, Grusch suffers a traumatic brain injury (TBI) during operations.

- Post-injury: He develops post-concussive syndrome, seizure disorder, and later PTSD tied to both the injury and the stress of intelligence work.

- 2018: Grusch experiences a mental health crisis. He voluntarily checks into an inpatient mental health facility for treatment related to his seizures, PTSD, and post-TBI symptoms.

• Post-2018: After treatment, he returns to work, passes background checks, and keeps his security clearances. He transitions to working at the National Geospatial-Intelligence Agency (NGA) and is also attached to the UAP Task Force.

• 2021–2022: Grusch begins officially submitting whistleblower complaints to the Inspector General, claiming knowledge of hidden UAP crash retrieval programs.[155]

• Mid-2023: He goes public in an interview with NewsNation and others, stating the U.S. government is hiding evidence of retrieved non-human craft. His previous medical history (the TBI, seizures, hospitalization) gets brought up in media coverage but he addresses it directly.

Grusch has been careful to frame his health struggles as unrelated to his credibility, and his supporters (including legal and military figures) have backed him up.

In a NewsNation interview, Grusch mentioned that he was later hospitalized in 2018 following a personal mental health crisis, partly tied to neurological complications from the injury — including the seizures. His medical issues were treated, and according to what's public, he has since recovered enough to pass regular security clearances, including the very high-level ones he needed to access classified UFO programs.

Here's a list of key government officials and figures who supported or validated David Grusch in different ways:

• Charles McCullough III: Former Inspector General of the Intelligence Community. He became Grusch's attorney and public representative, saying Grusch followed proper whistleblower procedures and was taken seriously.

• Christopher Mellon: Former Deputy Assistant Secretary of Defense for Intelligence. Mellon said Grusch is credible and encouraged Congress to listen to him seriously.

• Senator Kirsten Gillibrand: She helped lead bipartisan efforts to create whistleblower protections for UAP witnesses, and was instrumental in pushing for hearings after Grusch's claims.

• Congressman Tim Burchett: Vocal supporter of Grusch, saying the government has been lying about UAPs and that Grusch's testimony was brave.

• Congressman Mike Gallagher: Another figure who took Grusch's allegations seriously, calling for deeper investigation into alleged crash retrieval programs.

• Inspector General Thomas A. Monheim of the Intelligence Community (2022) found Grusch's whistleblower complaint to be "credible and urgent."

What is extremely intriguing is that other UAP whistleblowers with similar symptoms tied to UFO/UAP investigations have reported brain injuries, seizures, and neurological issues — not just Grusch! For example, Luis Elizondo hinted that pilots and intelligence officers exposed to

UAPs experienced brain damage, cognitive issues, and neurological symptoms. Gary Nolan, a Stanford professor, studied a group of intelligence officers and military personnel exposed to UAPs, finding unusual brain injuries, specifically in the caudate-putamen region of the brain, similar to effects seen in Havana Syndrome.

The *Havana Syndrome* refers to a set of mysterious medical symptoms first reported by U.S. and Canadian diplomats in Havana, Cuba, in late 2016. Individuals affected described hearing a strange noise or feeling a pressure sensation in their heads, followed by symptoms such as dizziness, headaches, nausea, memory loss, and cognitive difficulties.

Some scans even showed brain injuries similar to concussion trauma without any physical impact! As more cases were reported worldwide—including in China, Europe, and Washington, D.C.—the phenomenon grew increasingly alarming to U.S. intelligence and health agencies. The cause of Havana Syndrome has been fiercely debated.

Early theories suggested it could be the result of sonic weapons, microwave energy attacks, chemical agents, or even mass psychogenic illness. In 2020 and 2021, several U.S. intelligence reports indicated that directed energy—particularly pulsed radiofrequency or microwave radiation—was a plausible explanation. However, the origin of the energy, the technology behind it, and the entity responsible remain officially unresolved. Some believe foreign adversaries like Russia could be behind it, though hard evidence has been scarce or classified.

A fringe but growing theory ties Havana Syndrome to UFOs. Some researchers and experiencers suggest that the directed energy effects bear similarities to encounters with UAPs, where witnesses report electromagnetic interference, nausea, and brain trauma after close sightings. Additionally, certain intelligence figures connected to UAP investigations (similar those in the Pentagon's AATIP program) have hinted that unknown technologies, possibly non-human in origin, could interact with human biology in harmful ways, potentially including phenomena such as Havana Syndrome. The Havana Syndrome and UAP-related brain injuries do overlap in symptom type which includes headaches, seizures, balance issues, memory problems.[151]

Dr. Gary Nolan's research highlights this theory that UFOs can be causing neurological damage to the experiencers, pilots and those who get too close. Nolan confirmed studying over 100 cases of government/military people injured after supposed contact with UAPs or mystery phenomena. Many had scarring in the brain, specifically areas linked to higher cognitive functions. He said these people were often high-performing individuals (pilots, intelligence officers) who developed serious neurological symptoms after UAP incidents.[152]

Lou Elizondo stated that *some* people investigating UAPs — especially those up close — suffered physical symptoms, including seizures, radiation burns, and brain anomalies. He warned that UAPs might represent not just a flight risk but a "national security health risk" for personnel. It reveals that there is a high probability that what Grusch has experienced other UAP witnesses have as well. There is a connection with UAP exposure and these neurological injuries.[153]

So what did Nolan's research discover? He found in the brain scans of people (mostly intelligence officers, pilots, and special operators) who allegedly had close encounters with UAPs that there was significant damage to the Caudate-Putamen region of the brain.

Many of the subjects showed abnormalities in this region that is responsible for decision-making, planning, and movement control. The caudate nucleus was overdeveloped or scarred in many subjects, which is highly unusual. Nolan said this type of brain change is normally only seen in people with very high IQs or in advanced cognitive specialists — but here it was showing up with lesions, scarring, and sometimes inflammation.[154]

Just like Grusch, these similar symptoms we reported:

- Seizures
- Severe migraines
- Dizziness and loss of balance
- Memory loss and confusion
- Unusual sensory experiences (like hearing strange sounds or feeling pressure)

Some of the possible causes that Nolan has theorized include a focused electromagnetic energy or unknown directed energy. He also did not rule out that these injuries could result from close proximity to exotic propulsion systems — meaning technology that warps spacetime, and creates radiation fields. He went on to say "What I think is happening is that if you get too close to a UAP, it's going to screw up your brain tissue.". Essentially, it is not a coincidence, there is a connection.

This is very similar to symptoms reported by CIA and State Department personnel during the Havana Syndrome events. Bottom line, some researchers think it could be the same technology — possibly directed microwaves, radiation, or even exotic technology designed for surveillance, defense, or UAP propulsion. Are the UAP experiencers and those that are working within the intelligence community on special access programs and other projects associated with UFOs being targeted?

Dr. Gary Nolan is one notable figure who has publicly connected the dots between UFO encounters, brain injuries, seizures, and Havana Syndrome-esque symptoms. In multiple interviews (especially with The Debrief and The New York Post), Nolan directly said he believes some Havana Syndrome cases and some UAP encounters are linked by the same kind of neurological damage. Believe it or not, Nolan said the caudate/putamen injuries in UAP witnesses were identical in some cases to what doctors found in Havana Syndrome victims! "It looked like these people were being targeted with directed energy weapons or exotic propulsion technology, " said Nolan.

Luis Elizondo, during interviews and briefings has stated that UAP encounters caused measurable biological effects, including brain trauma and neurological illness, and compared it to Havana Syndrome. He warned Congress privately (according to reporting) that some pilots and investigators were being physically harmed by interactions with the phenomenon.

Christopher Mellon has hinted that the health effects seen in UAP witnesses are part of why the government began taking UFOs more seriously — especially when it mirrored Havana Syndrome symptoms. Then Senator and now Secretary of State Marco Rubio who was briefed on both Havana Syndrome and UAPs as a Senate Intelligence Committee member, suggested publicly that foreign tech or unknown tech could be responsible for both. Is this a unified front or just a coincidence that they are echoing the same connections?

David Grusch and Luis Elizondo have both either hinted at or been involved in projects related to consciousness, intuition, or remote perception, especially tied to national security. Elizondo spoke openly about consciousness possibly being a key part of how UAPs operate or interact with humans. Grusch referenced non-human intelligence (NHI) possibly manipulating or interacting with human perception — very similar to remote viewing concepts.

Both Grusch and Elizondo are tied to parts of the military or intelligence community that have covertly continued studying "psi" phenomena like remote viewing after the official shutdown of Project Stargate (the CIA's remote viewing program). They both worked at Space Force together. Currently Space Force has secretive divisions working on human-machine teaming, telepathic communication research, and extended perception studies, many of which grew out of remote viewing research.[156]

Personally, I have been warned by random people to avoid remote viewing as there are usually consequences for those who try and don't know what they are doing and could experience neurological damage. Many remote viewers and psychics (even back to Stargate Project) reported severe migraines, brain stress, and sometimes seizure-like episodes after intense sessions. Since I was young, I have always considered myself somewhat psychic and my wife can attest to this.

I have either through dreams or waking life have seen or penetrated a way to see into the future on certain events. 9/11 was one of them. I am certain that I have remote viewed without even realizing what I was doing. As I write this, I have been experiencing it more intently with information flowing and minutes later reality confirming it.

If UAPs are emitting exotic fields that interact directly with human consciousness, it would explain both neurological injuries (physical seizures, caudate/putamen scarring) and psychic episodes, hallucinations, or enhanced perception (remote viewing–like experiences).

My theory is that If UAPs involve directed energy plus consciousness interaction, then people similar to Grusch and Elizondo could be more sensitive or damaged neurologically due to close exposure. Their remote viewing abilities or intuitive skills might also be enhanced or disrupted as a result. Seizures could be a side effect of interacting with fields not fully understood, possibly linked to spacetime manipulation, electromagnetic fields, or nonlocal consciousness phenomena.

If you then add remote viewing and the brain's sensitivity to consciousness-driven phenomena into the equation, it actually strengthens the idea that whatever UAPs are, they're affecting both the body and the mind — and some people (Grusch, Elizondo, others tied to Space Force) are at the center of it.

Brain Regions Impacted by UAP Encounters: Paranormal Implications

Recent research by Dr. Garry Nolan at Stanford University has revealed that individuals who encountered UAPs exhibited unusual damage in two key areas of the brain: the caudate nucleus and the putamen. These structures, located deep within the brain, govern higher-order thinking, motor control, and decision-making processes.

Damage or overstimulation of these regions has been historically associated with:

- Seizures and neurological instability
- Enhanced intuitive abilities and altered perception
- Hallucinations and out-of-body experiences

Beyond UAP encounters, abnormalities in the caudate and putamen have appeared in individuals claiming experiences with:

- Poltergeist and haunting phenomena
- Remote viewing and psychic events
- Near-death experiences (NDEs)

This suggests that proximity to unknown fields — whether originating from advanced technology, consciousness anomalies, or energetic manifestations — may activate or disrupt the neural circuits tied to perception of non-ordinary realities. The overlap between brain trauma, heightened sensory awareness, and paranormal encounters presents a new frontier in both neuroscience and the study of the unexplained.

Due to the constructs of our brain, our physical manifestation and our soul containers keeping us plugged into this reality? Those "lucky" individuals who have been connected with UFOs and their close proximity being altered in such a way that they can now interact with other realities, dimensions or timelines? Now, what if there was a major shift about to occur recently in human consciousness but was somehow altered by the fear of widespread sickness? The timing of Covid-19 and the beginning of UFO disclosure could have a potential connection.

Major UFO/UAP Events Around COVID-19 Timeline:

1. 2019 "Pyramid UFOs" (USS Russell, West Coast, July 2019)

• Navy destroyers reported and filmed multiple green pyramid-shaped UAPs swarming military ships.

• This footage eventually leaked and was confirmed as real by the Pentagon in April 2021 — during COVID lockdowns.

• These craft performed maneuvers far beyond human capability (hovering, submerging into water, sudden 90-degree movements).

2. **2019 UAP Navy Leaks (NY Times)**

• In late 2019, more videos of Navy pilots encountering UAPs (including the infamous "Gimbal," "GoFast," and "FLIR" videos) were publicly acknowledged.

• This marked the first time in U.S. history that the Department of Defense officially admitted that the videos showed "unidentified aerial phenomena."

3. **2020 COVID Pandemic Begins — March 2020**

• Global shutdown happens just months after the Navy confirms the existence of UAP videos.

• Attention immediately shifts away from growing public awareness of UAPs toward survival, isolation, fear.

4. **April 27, 2020 — Official Pentagon UFO Video Release**

• In the middle of COVID lockdowns, the Pentagon officially declassifies the Navy UAP videos.

• This should have been a historic moment — "UFOs are real" — but global media barely focused on it due to COVID fear and political unrest.

5. **Summer 2021 — Preliminary UAP Report Delivered to Congress**

• For the first time, U.S. intelligence agencies deliver a formal report stating "UAPs are real, unexplained, and could pose a national security threat."

• Public reaction was muted due to ongoing pandemic chaos and vaccine debates.

Patterns You Can See:

• 2019–2021 was a massive acceleration of UFO disclosure — way more than at any time since Roswell.

• At the same time, the world experienced unprecedented fear, confusion, isolation, and neurological manipulation (social media, health anxieties).

• Potential Consciousness Awakening → Distracted or muted by pandemic response and fear-based mass programming.

The timing of the Covid epidemic along with significant UFO/UAP disclosure is highly suspicious if you're looking at this through the lens of controlling mass consciousness during a possible "awakening window." Now let's look at what is lining up for 2025.

As of April 2025, the U.S. has witnessed significant developments in Unidentified Anomalous Phenomena (UAP) disclosure, particularly under President Donald Trump's second term. Advocates are optimistic that this administration will prioritize both new and existing disclosure initiatives.

Key Developments in 2025:

• Executive Orders on Declassification: On January 23, 2025, President Trump signed an executive order mandating the declassification of records related to the assassinations of President John F. Kennedy, Senator Robert F. Kennedy, and Dr. Martin Luther King Jr. This move has spurred calls for similar transparency regarding UAP-related documents.

• Congressional Initiatives: Senators Chuck Schumer and Mike Rounds have introduced bipartisan legislation aimed at declassifying government records on UAPs, building upon previous efforts to increase transparency in this domain.

• Public Discourse and Media: The documentary The Age of Disclosure, featuring testimonies from 34 high-ranking officials, including Secretary of State Marco Rubio and former Director of National Intelligence James Clapper, premiered at SXSW 2025. The film alleges an 80-year cover-up concerning non-human intelligent life and has intensified public interest in UAP phenomena.

• Scientific Engagement: The All-domain Anomaly Resolution Office (AARO), established in 2022, continues to investigate UAP incidents. In a recent interview, AARO

Director Jon T. Kosloski emphasized the need for rigorous scientific analysis of unexplained aerial phenomena.

These recent developments suggest a growing momentum towards greater transparency and public engagement on UAP issues. Or are we headed towards another Disclosure disappointment?

Pilots and Physical/Mental Effects After 1952 UFO Encounters

Just like what the world is seeing now with these drone and orb incursions, military pilots in 1952 (and surrounding years) would also report physical and neurological effects after close encounters with UFOs, but at the time this was heavily suppressed or only whispered about.[159] Their accounts included the following symptoms:

- Severe headaches
- Confusion
- Temporary memory loss or distortion
- Difficulty concentrating
- Vertigo and disorientation

Not officially called "seizures" at the time (neurological terminology was different), but some symptoms mirror modern post-concussive syndrome or minor seizure-like brain disruptions (similar to David Grusch's issues after his TBI).

Pilot Captain Edward Ruppelt (head of Project Blue Book) and others noted strange behavioral changes among pilots who chased UFOs — confusion, missing time, even fear responses they couldn't rationalize. YES, pilots were experiencing neurological effects in the early contact era, but it was hushed to protect reputations and avoid panic.

Wouldn't this mind-altering effect on the pilots also indicate that this technology, UFOS or their propulsion system they are using isn't something that is human made? It is believed to be something beyond the circuitry or wiring of our neurological systems can handle? Seizures originate from abnormal electrical activity in the brain.

Normally, neurons or brain cells communicate in an organized way, but during a seizure, this activity becomes chaotic and excessive, causing changes in movement, behavior, feeling, or consciousness. A rewiring or reassembly that leads to unexpected seizures but maybe a new Is this an indicator that it is something interdimensional that is interacting with us because our brains cannot handle the output of the energy?

Epilepsy cases are still somewhat of a mystery. 50% of epilepsy cases are classified as having an unknown or idiopathic origin. What this means is that even after thorough testing (such as MRIs, EEGs, and genetic analysis), doctors can't find a clear cause or origin. Temporal lobe epilepsy, in particular, can cause intense spiritual or mystical experiences, deja vu, and altered states of consciousness.[157]

Scholars have speculated that biblical figures like Ezekiel, Elijah or even Paul after his Damascus vision of "bright light" may have had seizure-like experiences.[158] Philosophers like Socrates and mystics such as Joan of Arc and Saint Teresa of Avila have also been retrospectively considered by some researchers to possibly have temporal lobe epilepsy based on their descriptions of their visions and the voices that they heard. Here's a description of others and their experiences:

Ezekiel – Described vivid visions (seeing wheels within wheels) and unusual behaviors; often cited in discussions of neurological or epileptic visions.

Paul the Apostle – His dramatic conversion on the road to Damascus involved a sudden light, falling to the ground, and temporary blindness—sometimes compared to an epileptic seizure.

Joan of Arc – Reported hearing voices and seeing saints; some theories suggest epilepsy or auditory hallucinations.

Saint Teresa of Ávila – Had intense spiritual ecstasies and visions; some doctors have considered temporal lobe epilepsy as a possibility.

Mohammed – Early Islamic sources describe the Prophet experiencing shaking, sweating, and trancelike states during revelations—some scholars have controversially linked these to seizures.

Joseph Smith – Founder of Mormonism, reported visions of angels and divine beings; some modern theorists have questioned whether neurological factors played a role.

Ramanujan – The Indian mathematician credited his insights to visions from a goddess; some suggest these may have involved altered states or neurological inspiration.

Emmanuel Swedenborg – Philosopher and mystic who had visions of the spiritual world; modern analyses sometimes raise neurological explanations. Johnny Appleseed who was born in Leominster, Massachusetts was one of his initiates or followers.

Julian of Norwich – Had powerful visions during a serious illness; some have considered them seizure-related or hallucinatory.

Socrates – Spoke of a "daimonion" (inner voice) guiding him; some have proposed a neurological basis, though evidence is sparse.

Their visions and experiences, were they coming from another dimension? Nikola Tesla reportedly had vivid visualizations, extreme sensitivity to light and sound, and periods of trance-like focus. Some scholars have suggested temporal lobe epilepsy, synesthesia, or obsessive-compulsive traits, though Tesla himself said he could "see" inventions in his mind before building them. He also believed that he was communicating with aliens.

The US Navy's Unique Role With UAPs

The Navy emerges again and again in modern UAP interactions such as the USS Nimitz encounters (2004, "Tic Tac" UFOs) and theUSS Russell and USS Kidd incidents (2019–2020 pyramid UFOs). Just like Air Force pilots, Navy radar operators, sailors, and their pilots often describe a strange mental or emotional effect after encountering UFOs.[160] This overlaps with remote viewing, altered consciousness, and even the concept of being in the vicinity of portals:

○ Deep feelings of awe, disorientation, loss of time sense.

○ A sense of being "watched" or "communicated with" nonverbally.

○ Dreams and visions after close encounters.

There was one reported incident that is quite legendary and it is believed that Nikola Tesla was involved and helped develop the technology for naval invisibility using electromagnetic fields. This was ***The Philadelphia Experiment***. Tesla warned the military about the dangers before allegedly backing out. Some even believed he died under suspicious circumstances months before the experiment began in January 1943 because of it. With Tesla gone, the project would resume reportedly with the help of Albert Einstein. It went horribly wrong and if word got out that he was involved and would have tarnished his reputation as well as many others.

In Chapter 3, I introduced the idea that reality may be layered and that consciousness itself could function as a receiver or interface capable of interacting with dimensions beyond our ordinary perception. What begins to emerge is not a contradiction of it but a consequence of it. If UAPs and UFOs operate through fields that distort spacetime, electromagnetism, or perception itself, then exposure would not simply be visual or mechanical it would be cognitive and neurological.

The brain, acting as a biological filter, may be forced beyond its normal operating parameters when exposed to forces it did not evolve to process. Seen this way, seizures, hallucinations, precognitive experiences and altered states of awareness may not be failures of the brain, but overloading of the system. The same neural regions associated with intuition, planning, and perception may become both damaged and hyper-activated, producing a mix of injury and expanded awareness which leads to more encounters and experiences that cross the boundaries of cryptids, UAPs, the supernatural or the paranormal. This could very well be evidence of interaction with a system of reality that remains largely uncharted by modern science.

After Tesla's death it is believed that many of his ideas that became classified or disappeared were used in secret projects; examples would be early versions of the Star Wars defense program and the Manhattan Project.

Here are the most commonly claimed technologies that are believed to have been hidden or developed from Tesla's confiscated papers:

Death Ray ("Teleforce") – A particle beam weapon Tesla claimed could destroy aircraft from hundreds of miles away.

Wireless Energy Transmission – A system for transmitting free electricity through the air using the Earth's resonance.

Antigravity Technology – Supposed designs for flying discs or craft that manipulated gravity.

Time Travel or Dimensional Access – Theorists claim Tesla discovered ways to distort time and space using electromagnetic fields.

Free Energy Devices – Machines that could generate limitless energy without fuel, bypassing modern energy industries.

Mind Control or Frequency Manipulation – Use of resonant frequencies to influence thoughts, mood, or even health.

Invisibility Fields – Early versions of cloaking technology, often tied to the Philadelphia Experiment story.

Scalar Wave Technology – A controversial idea involving energy waves outside the standard electromagnetic spectrum.

Beyond mainstream science, Tesla's ideas have been taken up in alternative and holistic practices that focus on "energy healing," vibrational therapies, and the idea that everything, including thoughts and emotions, has a frequency that can influence our well-being.

We are now in late December of 2025 and there are rumors swirling about that President Trump, who has the lowest approval rating of any president, will be the one who will announce that aliens and UFOs are real. What's fascinating and many people don't know this is that his uncle, John G. Trump, who was an MIT electrical engineer, was tasked with reviewing Tesla's papers! The official statement was that Tesla's work was speculative and more philosophical rather than being actionable or usable. But if that is the case why were some of those documents never returned? Some of the materials remain classified and or missing to this day.

"I am strongly convinced that I have received signals from another world", Tesla publicly stated after experimenting with radio waves in Colorado Springs, he detected repeating, rhythmic signals. He openly discussed the idea of Non-Human intelligence and was potentially one of the first to do so. He believed that these were intentional signals coming from Mars, Venus or elsewhere and he vehemently believed that he had indeed made contact. Tesla believed in higher dimensions but that they couldn't be accessed by normal human senses.[161]

He believed in "ether" as a multidimensional all-pervading medium that carries energy, information and force and possibly structured in layers or dimensions.[161] Ether was replaced in modern physics as field theory. Tesla believed that reality was governed by frequency, resonance and vibration and that different states of matter or energy might exist outside of our perceptual range. His work would later evolve into wave-particle duality and higher-dimensional math using string theory. If his research was leaning into ideas about the fabric of our reality maybe he discovered something that wasn't intended to be shared.

"The day science begins to study non-physical phenomena, it will make more progress in one decade than in all previous centuries." - Nikola Tesla

Philadelphia Experiment Connection

The Philadelphia Experiment: Project Invisibility, published in 1979 by William Moore and Charles Berlitz explores the legend of a secret U.S. Navy experiment allegedly conducted in 1943 at the Philadelphia Naval Shipyard. The book claims that the Navy attempted to make the destroyer escort USS Eldridge invisible to radar using electromagnetic field manipulation. According to accounts, the experiment had unintended consequences, including teleportation, crew members becoming embedded in the ship's structure, severe psychological effects, and even time travel.[162]

Moore and Berlitz investigate various testimonies, most notably from Carlos Miguel Allende (or Carl Allen), who claimed to have witnessed the event and corresponded with researcher Morris K. Jessup. The book examines conspiracy theories, government cover-ups, and parallels with emerging physics concepts. Critics dismiss the story as a hoax, but the book fuels ongoing speculation about secret military experiments and unexplained phenomena.

Well, Morris Jessup ends up committing suicide under questionable circumstances. The mysterious man associated with the "Allende Letters" aka Carl Allen aka Carlos Miguel Allende also appears to have another alias or rather his real identity. Carl claims that he knows several people that are still alive who were associated with this classified project.[163]

The Philadelphia Experiment legend (1943) describes a U.S. Navy destroyer (USS Eldridge) supposedly rendered invisible or teleported to another port in Norfolk. Sailors on board suffered horrific side effects that included being "fused" into the ship's metal! They had severe disorientation, mental collapse, seizures, and some vanished without a trace.

Some of the key overlaps with this project include:

- Extreme electromagnetic field exposure.
- Manipulation of spacetime (wormhole or dimensional rupture).
- Human neurological breakdown under exposure.
- Memory loss, disorientation, and missing time.

Whenever humans (especially Navy personnel) interact too closely with exotic energy phenomena — whether UAP craft, dimensional portals, or field experiments. It triggers neurological effects, consciousness anomalies, and possible dimensional shifts that the body/mind cannot easily process.

I had the urge to reread the book. I read it when I was writing my second Monsterland book. I sensed that there might be an interesting connection on a local level. At about the halfway point on page 187, there it was. Like in the shadowy intelligence world of espionage, spies and the like with multiple identities, while author Charles Berlitz (who also wrote *The Bermuda Triangle*) was interviewing Dr Rinehart, the scientist proclaimed that there were several scientists that may have worked on this project or had been associated with it. He brings up the name of a legendary figure Professor Charles M. Allen. When asked if there might be a connection to Carl Allen, Rinehart said "No, I don't think so. Not CARL Allen ...Charles Allen. Charles Metcalf Allen was a professor in hydraulic engineering at Worcester Polytechnic Institute of Technology.[164] Was this the mysterious figure that was posing as someone else? Did he work for an agency tasked with making sure that the real information didn't get out?

Allen knew a great deal about ships, models, mines and so on. He is also believed to have been the one that Kent (his details) was integral in the discussions of how to produce the necessary electromagnetic field for experimental purposes by means of the principles of resonance or sound.

My daughter is attending and is currently an engineering student at WPI. I also learned that Ryan Graves, the US Navy pilot connected with the Tic Tac UFO incident happens to be from Gardner, Massachusetts, a stone's throw from Leominster State Forest aka Monsterland and he also happens to be a graduate of none other but WPI. This was shared with me at a recent lecture I gave where a member of the audience said that his son and Ryan were friends and he thought I would be interested to know that.

I reached out to Ryan and heard back from him in a private message on X. Prior to this, before my daughter started school in the fall, I discovered an unlisted scholarship program by the CIA at WPI...that is invite only. They are actively recruiting at the University and the new President is former NASA. We are dealing with some of the youngest and brightest minds in New England and all these acronyms have another connection to UFOs or UAPs.

A few years ago, for six hours I interviewed a family that lived in Westminster near Leominster State Forest. They had reached out to me about having me come by so they could discuss recent developments and encounters that they have been having at the house with Bigfoot, orbs and other strange things. What was intriguing is that the mother shared with me that her husband, who has now passed, was a scientist with 3 PHDs and several near death experiences that gave him insight into other realms. The family would experience all the strange phenomena from Bigfoot, UFOs, shadow people and orbs. She said that he worked with Einstein on projects focused on dimensions and other classified projects. They were moved here to be close to FT. Devens. Who was he?

This UFO/UAP enigma exposing itself at an exponential level has been happening since at least since the early 1940s, exploded again in 1952, and is happening again now under different names.What we are seeing now in the skies, what pilots and whistleblowers are reported to Congress and the Senate isn't new but it is coming to a place where I think more will be revealed soon.

If disclosure is truly happening as many believe then it is not just a revelation about advanced craft and retrieval programs but a reveal about who we really are, what are limits are when it comes to our biological makeup and the vulnerability or manipulation of our consciousness. I think it is becoming clear that it is not a question whether something is interacting with us but how long it has been happening and what's the cost of that interaction.

Chapter Thirteen

MAINE MONSTERS

One of my favorite places on the planet is the state of Maine. To give you some perspective in regards to how significantly large and blanketed in forests, Alaska possesses the largest total forest area in the United States, with approximately 129 million acres of forested land, encompassing both boreal and coasts temperate rainforests. On the other hand, Maine with 17.5 million acres has the highest percentage of forest cover among the states with about 89.46% of its land area covered in forests. Of this, approximately 86% is classified as timberland.

The earliest known report of a Bigfoot-like creature or "wild man" in Maine dates back to the early 1800's, though oral traditions from indigenous tribes predate that. One of the first documented sightings was in 1855, in Waldoboro, Maine when a newspaper, the *St. Albans Weekly Messenger* published a story on February 8th.[165] The report that comes out of Waldo County described the experience of J.W. McHenry, who while chopping wood near his home, heard a terrifying scream and saw a hairy, man-like figure emerge from the woods where the sound was coming from.

The Wabanaki tribes also have older legends of large, hairy forest beings, sometimes described as forest guardians or spirit beings, similar to Sasquatch. These accounts are often very spiritual in nature but align with later settler reports of similar creatures. Are they the same being but with the tribes having more knowledge of what they really are and that they are interdimensional or spiritual?

We often hear positive stories with a lack of attacks, especially fatalities when associated with a Sasquatch, but in 1886 an article appeared in the *Waterville Sentinel* describing a chilling encounter with a 10-foot-tall "wild man" in the woods of Maine. According to the account, a French-Canadian hunter reported that while camping about 100 miles north of Moosehead Lake, he and his companions were attacked by a massive, hairy creature with 7-foot-long arms![166]

The creature allegedly killed one of the men, prompting the others to gather up a dozen men to hunt it down and shoot it. It was reportedly "killed". The article also states that this was a part of a broader pattern of "wild man" or Bigfoot-like sightings in Maine during that era.

During August and another visit to northern Maine in September of 2024, along with twin brothers Travis and Toby Hartford of Cryptic Paranormal Adventures of New England, we experienced not only Bigfoot but UFO activity. The location sits in the middle of the above mentioned reports of the late 1800's. The twins put together a Squatchin' and Skywatchin' event on a remote 300-acre private property.[167]

We limited the attendees to thirty people for each event. The location remained secret until a few days before to avoid any encroachment or other human contamination. The property was selected due to recent activity emanating out of the adjacent swamp. This would be my son Noah's first organized Bigfoot hiking event with me. It didn't disappoint.

The plan was to set up tents at the field for an overnight hike looking for Bigfoot evidence or even better capture something on camera as well. Once night was in full effect, we would then sit in front of the campfire and watch the night sky for anomalous activity. Once camp was set up and burgers were on the grill, Mike Vashon of the *Maine Bigfoot Foundation* and I did a short presentation of what the game plan for the evening would be and what to look for.

The idea was to break up the thirty people into three groups of ten, each group would be led by myself, Mike, Toby and Travis. The areas of focus were within a short walking distance of camp. Once sunset hit we would make our way into the different spots.

Mike shared an incredible story about a hunter in Maine that he had been working with. Mike's uncle was a very prominent and well-known researcher in Maine. His uncle had garnered a lifetime of information, experiences and stories from hunters and residents about their tight-lipped experiences they were initially unwilling to share, eventually enriching his nephew with the knowledge and handing the baton to Mike to continue with the work.

Vashon shared some of the cases he was working on up in Vacationland. One story would be monumental. "There is a hunter that I have been talking to that has seen a sasquatch in his favorite hunting spot. It seemed that every time that he would venture into this area it would make itself known but kept its distance. With each visit, it would get a little bit closer. The hunter would use an ATV to access the very remote hunting location, leave the vehicle on the trail and he would walk into the forest to start stalking deer.

After several hours of hunting and no luck, he returned to the location of his parked ATV. As he got closer, he noticed something yellow leaning against one of the tires. He knelt down and was astonished to find a bundle of yellow flowers wrapped in a bouquet held together by some sort of vine, leaning against the tire. He couldn't believe it. He looked around, there were no tracks leading to his ATV. There was no one there that he would see. Was he just given flowers from a Bigfoot?"

We digested the story with awe. We then split into the three groups. I was at the most active area of the property. I had the swamp for all three of the runs. A few months earlier a series of 17-inch footprints with visible toes and a six foot stride were discovered at the edge of the swamp. Mike had his area to the right of the swamp along a dirt trail and deeper into the woods.

Travis and Toby took the opposite side of the property by the old homestead. Toby also shared with me that he and Mike were at the location a few years ago and Mike had spotted tree limbs snapped at the 10 and 11 feet height mark! We would do about an hour, head back to the tent area and debrief one another. We would take 15-20 minutes to regroup and head back into our designated areas.

Almost as if someone or something was listening when Mike and I were talking with the group before heading out, on Mike's second lap and returning to the camp, darkness had fallen fast so there was no visible light. Making their way back to the tents with the aid of flashlights, Mike leading the group stopped dead in his tracks as he noticed something yellow sitting in the middle of the trail as he was getting close to camp. Mike approached with a slower pace and right there in the middle of the trail, it wasn't there before, a bouquet of yellow flowers wrapped and held together with what looked like another plant or vine to keep the flowers intact.

When we all got back to camp we could hear the group almost giddy with excitement regarding their find. Mike, ever the professional, using rubber gloves with hopes of extracting some eDNA from the flowers retrieved them and sternly asked the rest of us if one of us had put the flowers there. "Seriously, I need to know right now, did any of you do this?" We soon realized that no one was back at camp. "Well, if that's true then we just got gifted these" It quickly dawned on us all that only a few hours earlier Mike had mentioned the yellow flowers being gifted to the hunter. It was no coincidence and the energy of the group escalated with excitement. It was about to get even better.

What made these events impossible to dismiss was the apparent intelligence behind them. The gifting of yellow flowers mirrored a previously documented Maine hunter encounter almost precisely, suggesting memory, repetition, and intentional symbolism rather than random behavior. We weren't in the same location that the hunter was in so that was amazing.

I believe that they are listening and can understand what we are saying if they have the ability to mindspeak, language is irrelevant. The flower delivery was interpreted as friendly and a connection forming event that night that had thirty people so excited that we were being interacted with not only flowers, but boulders being thrown into the water instantly on command.

We had a UFO experience as well as what some might call cloaking or a paranormal experience involving a shadowlike figure. When viewed through this lens, the phenomenon behaves less like an undiscovered animal and more like a responsive presence that is capable of observation, communication, anticipation, and engagement.

The first event, while we were at the swamp investigating with one of the groups of ten, Toby Hartford was with our group and had his FLIR camera rolling. He noticed what resembled a large head and shoulders of something submerged to about chest height in the water. It seemed to be positioned and facing the camera straight on. As he was filming this, it looked as if the "figure" was moving a little bit to the side and back. We would then return about an hour later to the swamp with another group and after a few minutes as we surveyed the area with our toes as close to the swamp's edge as possible we heard some movement in the water. Then an enormous splash of what sounded like a big rock crashed into the swamp only feet away from where we were standing. People screamed.

My adrenaline starts to pump and Noah gets a little bit closer to me. After a few minutes I politely asked if it could throw another rock in the water for us. And in no time, another rock was thrown. This time it was bigger sounding and even closer. It sounded similar to a boulder. The after effects of the impact were alarming. I will often describe the equivalent of when you see someone doing a cannonball into a pool and the sound of the water that was pushed into the air above comes raining down after the impact. The sound of the rain lasted for what was almost 3 to 4 seconds! Did it throw something or did it launch itself into the water? More screams. It gave me chills that it responded.

The activity at the swamp did not occur in isolation. Physical interactions, footprints, rock throws, water displacement which were immediately followed by aerial anomalies above the same location, witnessed by multiple participants simultaneously. This convergence of terrestrial, aquatic, and aerial phenomena reflects a recurring pattern documented globally, where Sasquatch-like encounters, paranormal happenings and unidentified aerial phenomena appear to co-manifest within the same geographic and temporal window.

The action at the swamp seemed to come to a standstill and we got the call on the walkies to head back to the tents. As we were walking back, still vibrating from the interactions, we could see the stars and sky so clearly above us. Movement caught everyone's attention when a formation of white orbs or spherical lights were flying above us from left to right. Noah quickly pulled out a green laser light and pointed it at one of the lights.[168]

The orb that he hit stopped in mid-air as the other four continued on their route. The orb motionless for a beat quickly changed its trajectory and at a higher rate of speed went back in the direction in which it came, leaving the others behind. We all just looked at each other "Wow!"

When we got back to the tents we shared what transpired. We heard from another attendee that they were seeing orbs moving through the woods at the time we had rocks being thrown! On the third and final return to the swamp in the field to the left I noticed what looked like an upright shadow, about six feet tall closer to the tree line but walking in the other direction. When I put my flashlight on it, I could still see it but the shadow, still moving, was somewhat translucent. What did I do? I ran towards it! But when I did this, it vanished and I could feel a bit of a chill or temperature change when I got to its location.[169]

I quickly rejoined the group and found my son and got in line next to him. He then told me what happened when I ran over to the shadow. The shadow in a heartbeat was standing next to Noah and walking along with him and the others on the trail. Noah thought it was my buddy Keith who was walking next to him in my place. Keith thought someone was standing behind him. When they both turned to look there was no one there! The shadow had joined the group!

The shadow figure's behavior where it was partially visible, mobile, and capable of appearing within a group without producing sound or physical disturbance. This aligns closely with reports of liminal or phase-shifted entities described in both Indigenous traditions and modern anomalous research. Its interaction suggested proximity without threat, presence without permanence, and visibility governed by conditions not fully understood. Such encounters challenge the assumption that all non-human intelligence must exist fully within our physical spectrum.

Noah was excited but clearly scared at the same time when we finally settled in for the night. This was his first event and we had another one here in Maine at this location about a month away. On the drive out in the morning with very little sleep, Noah said, "I don't know how we are going to top that one! That was insane!" When we returned in September, the event hadn't even started when two girls decided to explore the swamp while they still had daylight because they had heard about all the activity from the last event.

To everyone's shock, they discovered a series of thirty footprints coming out of the swamp. This wasn't just one footprint that could be a bear or something else but an entire trackway. These were enormous! They were 20 inches long and 9 inches wide! A calculation that is typically used when calculating height from the size of a footprint which estimates that the size of the creature was 11 feet![170]

This height of the creature being eleven feet falls in line with the old reports from the state of Maine. The height sounds ridiculous until you see the prints for yourself. There was another surprise, Mike had casted one of the best prints out of the group and when we were finally able to analyze it once it was dried, night had quickly come upon us. With the use of flashlight, empty pill bottles and several tweezers Mike and a member of his team discovered a series of white hairs on the bottom of the casted print! There were a lot of them and an ample amount of hairs to submit for a sample.

Mike's niece is a forensics scientist with the New Jersey police so he sent several of these hair samples along for her to test. We recently learned that the eDNA results came back as "partially human" not Homo Sapien, but something that had partial human DNA! Was this the infamous "Wild Man" of the Woods?

In Christopher Packard's *Mythical Creatures of Maine*, there were many stories from the Wabanaki tribe that resembled Bigfoot. They described these cannibal giants known as ***Chenoo*** or ***Kiwakw*** which are six to ten foot tall hair hominoids. What makes them unique is that they are very dangerous and known to be man-eaters and more human than animal! Were these the casts of the Chenoo of Native American folklore?[171]

Physical evidence emerged alongside these encounters, but it did not simplify the mystery—it deepened it. Footprints exceeding twenty inches in length, hair samples producing partial human markers, and repeated structural manipulation of trees point toward something that resists easy biological classification. Rather than forcing these findings into a single explanatory box, they may be better understood as indicators of an intelligence operating at the boundary between physical and non-physical states.[172]

My son Noah and I returned to Maine again a few weeks later for another event at the same property. We were joined again by the Maine Bigfoot Foundation and the original group of researchers, along with a whole new group of people looking to have an experience. Similar to last time we arrived around 4PM with plenty of sunlight lingering for a few more hours. We heard about some interesting structures that were found along with some deep impressions.

Mike and Kevin had casted two separate prints that were in the immediate area where structures were located, an X as well as another tree with no branches bent into a 90 degree angle at about the 4 foot mark. While they were setting the Plaster of Paris into one of the prints, Kevin, from the corner of his eye witnessed a glimpse of a tall dark shadow running between two trees. When he started to walk out of the area he heard a very deep growl.

Mike while setting the material into the second cast would also witness one cross by the area where the structure was and the tracks. Mike estimated its size to be about 9 feet tall! Two visual sightings and two prints that would eventually be revealed to be over 22 inches and the other 23 inches long! Absolutely enormous.

Once we set up camp, Noah and I surveyed the swamp area again. There was a pile of cut trees, chopped up for fire wood and stacked. We noticed a tree, no limbs bent over at a 90 degree angle. The tree was about a foot wide and on the opposite side where the footprints were discovered and drying. We confirmed that the tree was identical to the one we had over at the swamp. They were both pointing in the same direction. There were two other trees forming an X. The trees used were similar to the ones shaped at 90 degrees - birch trees.

Once nightfall had hit we set foot into the swamp again with the first group. We had a total of four rocks being thrown into the swamp from the other side at us during this first hour, and another four thrown during the second hour! Mike and his team were to the right of us closer to camp in the woods in the area where the casts were located and they were getting heavy footfall, stomping and growling.

When we were returning to camp before the second pass and we arrived at our tents there was a long bird-like whistle from the woods that we were just in and we had a response come from the other side of where we were camped. It was a series of three chirp sounding responses to the first long whistle.

It has been reported that sasquatch will use whistles or animal mimicry to communicate with other members of their group. We discussed how it didn't sound like a bird at all. The sound seemed fabricated. This was late at night and I got the sense that there were two that were communicating between each other. We couldn't get a visual on them unfortunately, but I had that feeling in the pit of my stomach that it was "them."

Maine did not offer answers but it added to the layers, confirmed the activity is happening in the same place and connecting with individuals who have already had an interaction. A connection was already made. They had our number.

The experiences in Maine have offered continuity, reminding us that whatever has walked these forests for centuries has never truly left.

They are still here.

(Fig. 13.1 Author with cast on a private property for the Skywatchin' and Squatchin' event, Somewhere, Maine with a 20 and ¾ inch foot print / 9 inches across. Photo by Noah LeBlanc)

Chapter Fourteen

THE GREEN MAN OF ENGLAND AND SASQUATCH OF NORTH AMERICA MYSTERIES INTERWOVEN

I went to England for the first time in 2024 after leaving the show Expedition Bigfoot. I participated in a documentary called My Bigfoot Life starring Danile Lee Barnett. It was an eye opening trip. What led me across the pond was his discovery of an 18-inch footprint, he casted it and then DNA evidence was gathered from that footprint which led to the discovery of "Great Ape and Old World Monkey" DNA pulled from his cast. There appears to be a potential link or connection between the Green Man of England and Sasquatch of North America.

There are some intriguing similarities that transcend geographical boundaries. Throughout the world there is a creature that is reminiscent of bigfoot sharing the same traits and features. Some examples are the Yowie of Australia, the Yeren of China, the Almasty of Russia and so on. Are they permutations of the same thing?[174]

The connection between these beings might not just be folklore but part of a larger, more complex phenomenon involving elementals, spiritual entities, or even interdimensional beings. I am not satisfied with just collecting and casting footprints. Where are the bones? Where are the bodies?[175] This theory tends to carry more weight based on eyewitness descriptions and the associated high strangeness that seems to envelop these beings as well as the stories from the ancients, not to mention those of the indigenous peoples throughout the world.

The Green Man and Sasquatch, while appearing in different cultural contexts, exhibit remarkable parallels in their descriptions and the phenomena associated with them. The Green Man, often depicted in carvings and folklore as a man's face peering out from leaves, is seen as a symbol of nature's vitality and regenerative power. Sasquatch, known for its massive, hairy humanoid form, is similarly intertwined with the natural world, often seen as a guardian of the forests. They both contain human or man-like qualities, as well as spiritual significance that evoke a potential link to our past history of how we became who we are today. Is it just evolution or did we get a little help from some friends?

The Native American tribes across North America have rich traditions and stories about Sasquatch. The Sts'ailes people of British Columbia call it "Sásq'ets," a being that is deeply respected and considered a protector of the forest. The Yurok tribe in California speaks of "Oma," a creature that lives in the remote mountain regions and possesses supernatural abilities. These legends often describe Sasquatch as a spiritual entity, capable of great strength and wisdom. That spiritual definition fits nicely into the dimensional bucket doesn't it?[177] [178]

The Miwok people of Northern California have a tradition of a "hairy man" who lives in the wilderness and watches over the land. Similarly, the Hopi tribe speaks of a "giant" that roams the forests and mountains, serving as a guardian spirit. These stories, passed down through generations, emphasize the importance of respecting nature and living in harmony with the environment.[179]

Historical accounts of Sasquatch date back to the early explorers and settlers. In 1811, British explorer David Thompson reported finding large footprints in the snow near Jasper, Alberta. In the mid-1800s, settler accounts began to emerge, describing encounters with a large, hairy creature in the wilderness.[180]

In England, another figure could be associated with or be one of the same as the Green Man but could be something different altogether. The Wild Man of the Woods or Woodwose is a figure dressed in green leafy clothing. He is seen carrying a staff, a club, a bow and arrows could be elements that have survived from ancient mythology. This image is typically seen hanging in signs outside pubs that have some sort of reference to the Green Man. He can oftentimes be seen, sometimes covered in fur or hair, carved into the stonework of medieval churches.[181]

Both of the New England and England entities of the forests have been linked to mysterious lights. There are reports of strange orbs, balls of fire or orange orbs seen in the vicinity of these sightings. That can't just be happenstance. I don't believe in coincidences.

Similarly, the Green Man has been associated with unexplained lights and mysterious occurrences in the forests of England. These commonalities raise intriguing questions about the nature of these beings and their potential connection to other paranormal phenomena, such as UFOs. The idea that Bigfoot may well be an elemental being further strengthens this connection across the pond where those lights are often described as the fae or fairies.[182]

Why we don't have bones or bodies but sightings on videos or photos, physical impressions left in the ground, massive tree structures and other anomalies may shift back to the idea that they very well could be coming in from another dimension. Are these the higher beings which some may consider to be labeled angels? Could they have some sort of involvement in our spiritual development?

There was a story that was shared by the Yakima tribe. They insist that this isn't some myth or folklore that has been passed down by a real event. They shared this with J.W. Burns who was working with the tribe teaching them to write and read English. They trusted this white man sharing something that has only been told within the tribe for hundreds of years. The story goes that a tall, hairy humanoid being with glowing orange/reddish eyes.

It came to live with the tribe. It had the ability to speak with all the members, not verbally but through telepathy. When any of the tribal members became ill or sick this being had the ability to heal them. It lived with the tribe for many years but it communicated to them that its time was nearing to a close and that it needed to be brought to a specific location. The tribe obliged and as it lay there taking its last breath they watched in amazement once it passed away as a glowing light or UFO descended from the sky and came down and retried the body and then flew away. So based on this story it indicates that this being had a shelflife and a limited time here on this plane.[183]

Are the Green Man and Sasquatch spiritual guardians, remnants of ancient elemental forces, or interdimensional travelers crossing the boundaries of our reality?[176] Whatever they are, people are having more experiences in the present day and it doesn't seem to have any intention of slowing down anytime soon.

The Salish people of the Pacific Northwest describe Sasq'ets or Sasquatch as a spiritually powerful forest being and not just a flesh and blood physical creature proclaiming that they move between the physical and spiritual world. Ethnographers such as Wilson Duff and Diamond Jenness have documented narratives in which these beings exhibit the ability to appear or disappear, a phenomenon attributed to their status as spirit persons rather than being strictly animals.[177]

The elders describe them as people of the trees who step between worlds through the cedar forests. The cedar tree is considered sacred and believed to hold spiritual pathways. The Hoopa Valley in northern California consider these beings as forest guardians that sometimes "walk into the trees" or "become part of the woods". Cherokee legends of Nun'Yun'Wi and similar forest beings often describe their movement through "shadow places" or concealed forest passages which in modern day we would consider to be portals.[184]

It is important to clarify why the comparison to the Green Man to Sasquatch matters within the broader arc of this book. What emerges from indigenous testimony, modern encounters, and recurring symbolic patterns is not a collection of isolated myths, but a consistent intelligence associated with forest, liminal spaces, and moments of heightened awareness.

Whether named Sasquatch in North American or Green Man in England, these figures seem to occupy the same role as guardians of the land, mediators between worlds, and reminders that consciousness may be embedded within nature itself. It is through this lens, not as folklore alone, but as a recurring interface between humanity and a living environment that the Green Man must now be examined.

The Green Man of England

Origins and Historical Background

The Green Man, a figure found carved in wood and stone in churches, abbeys, and cathedrals across Europe, particularly in England, is a representation of a face surrounded by or made from leaves. This imagery dates back to at least the 2nd century, with the Green Man often linked to ancient pagan beliefs and the cycle of life, death, and rebirth associated with nature.[185]

The Green Man's origins are shrouded in mystery, with his image appearing in various forms across different cultures. Some of the earliest depictions can be traced back to the Roman god Bacchus and the Celtic deity Cernunnos, both of whom were associated with nature and fertility. The Green Man's face, often surrounded by foliage, suggests a deep connection to the cycles of growth and decay in the natural world.[186] [187]

In medieval Europe, the Green Man was incorporated into Christian architecture, appearing in the form of grotesques and decorative carvings in churches and cathedrals. This blending of pagan and Christian symbolism suggests a cultural attempt to harmonize the old beliefs with the new. The Green Man became a representation of the eternal cycle of life, death, and rebirth, reflecting humanity's dependence on and reverence for nature.

In the 1930s, Mr. Griffith, the Vicar of Saint Joe's Church in Monmouthshire and a folklore enthusiast, introduced Julia, Lady Raglan, to an image that intrigued him: "a man's face, with oak leaves growing from the mouth and ears, completely encircling the head." Until that time, the foliate head had no specific name. Mr. Griffith believed the image symbolized the spirit of inspiration, but Lady Raglan interpreted it differently. She was convinced it represented a man rather than a spirit and decided to name it the "Green Man." Her fascination with this iconic figure led her to publish an article in the Folklore Journal of 1939, where she detailed her interpretation.

The symbolism of the Green Man extends beyond Europe. In ancient Mesopotamia, a similar figure known as the "Master of Animals" can be found, representing a deity who controlled and protected wildlife. In Islamic art, the "Green Man" appears as Al-Khidr, a wise and immortal figure associated with water and vegetation. These cross-cultural parallels highlight the universal nature of the Green Man as a symbol of life, fertility, and the natural world.

Sightings and Modern Encounters

While primarily a symbol, there have been numerous reported sightings of the Green Man, particularly in rural England. Witnesses often describe a humanoid figure covered in foliage, blending seamlessly with the forest environment. These encounters are typically brief and occur in dense woodland areas where the figure is seen moving among the trees before disappearing.

In addition to the 1962 sighting in the Forest of Dean, there are other notable encounters. In the 1980s, a series of sightings in Epping Forest described a similar figure. Witnesses reported seeing a man-like entity with skin and hair resembling moss and leaves. These sightings often occurred at dusk or dawn, times traditionally associated with supernatural activity.

One particularly interesting account comes from the New Forest in Hampshire. In the early 1990s, a group of hikers reported seeing a figure that appeared to be part of the forest itself. As they approached, the figure seemed to dissolve into the trees. This sighting was accompanied by an eerie silence and a sense of being watched, common elements in Green Man and Bigfoot encounters.[188]

In another case from the early 2000s, a local historian in Cornwall reported an encounter with a Green Man-like entity near an ancient oak grove. The figure, described as a tall man with bark-like skin and eyes that glowed green, appeared during a midsummer ritual being performed by a group of pagans. This sighting was particularly notable for its context, as it occurred during a traditional ceremony honoring nature spirits.

There have also been reports of the Green Man appearing in modern urban settings, often near parks or areas with significant greenery. In 2010, a resident of London claimed to have seen a figure resembling the Green Man in Hyde Park. The witness described the figure as a tall, green-skinned man who seemed to be observing the park's visitors before vanishing into the trees. This sighting, though unusual, suggests that the Green Man's presence may not be confined to rural areas.

The Role of Sacred Groves

Sacred groves, areas of woodland revered and protected for their spiritual significance, have long been associated with the Green Man. These groves, often remnants of ancient forests, are seen as places where the natural and supernatural worlds intersect. They serve as sanctuaries for wildlife and as sites for rituals and ceremonies honoring nature spirits.

In the British Isles, sacred groves were integral to the spiritual practices of the Druids, who performed ceremonies and rites in these natural temples. The association of the Green Man with these sacred spaces suggests a deep connection to the ancient traditions that revered the forest as a source of life and spiritual power.[189]

Similarly, there are Bigfoot reports emanating out of areas considered to be old-growth forests, sacred or power place locations that some may consider containing some sort of a portal or doorway to another dimension or place. And some locations simply because they potentially have healing qualities. The Abenaki tribe, believes that Bigfoot "The Forest Wanderer" also known as *Kchi Awas* protects sacred areas such as Brunswick Springs in Vermont and Bradford Springs in New England.

Sasquatch is known to many indigenous people as protectors of the forest. These beings go to these special springs for healing.In his book "The First Vermonters," former University of Vermont anthropologist William Haviland observes that the Abenaki recounted stories of "The Forest Wanderer," a giant humanoid creature that occasionally left footprints behind. These accounts parallel Algonquin tales of the windigo or "giant cannibalistic man."

Ray Owen, son of a Dakota spiritual leader from Prairie Island Reservation in Minnesota, told a reporter from (the) Red Wing (Minnesota) Republican Eagle, "They exist in another dimension from us, but can appear in this dimension whenever they have a reason to. See, it's like there are many levels, many dimensions. When our time in this one is finished, we move on to the next, but the Big Man can go between."[190]

"The Big Man comes from God. He's our big brother, kind of looks out for us. Two years ago, we were going downhill, really self-destructive. We needed a sign to put us back on track, and that's why the Big Man appeared". The Hopi, the Sioux, the Iroquois, and the Northern Athabascan -- Bigfoot is seen more as a sort of supernatural or spirit being, whose appearance to humans is always meant to convey some kind of message."

The Green Man as an Elemental Being

The Green Man's association with nature and the changing seasons has led some to speculate that he is an elemental being, a guardian spirit of the forest. Elementals, according to various mystical traditions, are entities that embody the fundamental forces of nature: earth, air, fire, and water.

In many accounts, the Green Man is seen in areas with significant ecological or historical importance. These sightings often coincide with periods of environmental stress or change, suggesting that the Green Man may appear to remind humanity of its connection to nature and the importance of preserving the natural world.

Elemental beings much like the Green Man are often described in esoteric traditions as having control over natural elements and forces. The Green Man, specifically, is thought to embody the spirit of the earth and the forest. His manifestations are said to coincide with the turning of the seasons, appearing more frequently during the equinoxes and solstices, times of significant natural change. In a lot of these carvings and depictions, there is plant matter emerging from his lips. This disgorging of foliage from the Green Man's mouth may represent God's creation of the natural world as told in the book of Genesis.

The concept of the Green Man as an elemental being is reinforced by his presence in various mythologies and folklore. In Celtic traditions, the Green Man is associated with the Green Knight, a mysterious figure who tests the hero's courage and virtue. In medieval alchemy, the Green Man represents the alchemical process of transformation and renewal, symbolizing the eternal cycle of growth and decay. Green is the color associated with the natural world and is considered to be the fairy color, particularly in Celtic countries. In Lancashire, fairies are called "Greenies" for the green clothes they are said to wear.[191]

There are also reports of telepathic communication with Sasquatch, where witnesses describe receiving messages or impressions from the creature without any vocal interaction. I have experienced this first hand as I described earlier in the book and have witnessed others experiencing this too. This phenomenon, known as "mind-speak," is often but not always reported in conjunction with sightings of other paranormal phenomena, such as orbs or strange lights.[192]

Mind-speak is the ability for these beings to communicate with you via telepathy. It is believed that they can read your intentions when you walk into the forest and like Dr. Rebecca Foster states when she received a message from the shadow being that it was going to run. These mind-speak communications can be thoughts or words, images, sounds and feelings transmitted to the person. They are still communicating with her even though she is back in Minnesota, asking her to come return to Ohio.

Mind-speak is considered a part of the "woo" when it comes to sasquatch. This piece of the puzzle is usually shunned unless someone has had a direct personal experience. It sounds too much like science fiction as they are still trying to grasp the creature's existence in the first place and then you throw this gem at them. Along with the cloaking ability that I have also personally witnessed which was filmed on the docu-series *Expedition Bigfoot*, these are elements that most researchers avoid due to ostracizing and ridicule, but I believe these "elements" hold the key to what they really are.

When it comes to telepathy animals can't do this, as far as we know, but humans can. We know this now. Telepathy is also reported with UFO encounters when abductees or experiencers describe the communication method between the pilots or occupants of these craft during sightings or abduction memories.

The connection between the Green Man, Sasquatch, and elemental beings is further supported by various spiritual practices and rituals that honor these entities. In pagan traditions, rituals to honor the Green Man often involve offerings of flowers, fruits, and other natural elements. These rituals, typically performed during the equinoxes and solstices, seek to align participants with the rhythms of nature and the spirit of the Green Man. This falls in line with "gifting" another component of interaction with Sasquatch that many people report as one of the entry points kicking off the relationship into the stranger side of things like seeing shadows and orbs at research areas or even in the researcher's home.[193]

Similarly, Indigenous ceremonies that honor Sasquatch often involve rituals of purification, prayer, and offerings to the spirits of the land. These ceremonies seek to establish a harmonious relationship with Sasquatch and other spiritual beings, recognizing their role as guardians of the natural world.

I have been told by several tribal members from different tribes that when they perform their sacred ceremonies the Sasquatch will show up and participate in them. They will appear in physical form, walk into the space, sit down next to them and they will sing their songs and speak in their native tongue. I am told that they also have other visitors who are more ethereal or made out of light that will manifest as well during these ceremonies. I hope to one day be a part of one of these. So are the elementals? Are the Green Man and Bigfoot spiritual beings bound to protect nature?

3 BIRDS Case Study: Synchronicity and Land Intelligence

Synchronicity is always tugging at my sleeve.[194] While I was in the midst of writing this chapter about the similarities of the Green Man and Sasquatch, New England and England, I had to stop to bring my daughter to school. While we were driving there, a red-tailed hawk dropped down from one of the trees along the highway, swooping down and flying in front of my car. This happens so often it's expected and weird at the same time. I have identified the redtail hawk as one of my spirit animals. It went from left to right. I tell my daughter Aysia, "This appears to happen more often than not and it is always directly in front of my car or over my car where I can see it, not behind or ahead of me. The timing is always uncanny" I dropped her off and then headed back home so I could finish up the chapter.

While I was writing, I was connected and in the zone. I was in the flow state and gaining momentum getting close to finishing the book. Instantly thinking about heading back home to finish the chapter, I see another red-tailed hawk drop down from the left again and fly in front of my car flying across my vision and the road left to the right. "What the?" I say out loud to myself. About a few miles later I am now scanning the sky and more aware of my surroundings. Keeping my eyes on the road but glancing around and I notice high up in one of the trees to the left of my a vertical shadow perched on one of the many branches.

It was the size of the silhouette that caught my attention, extremely large. As I got closer I could see that it was a bird of some kind, maybe even an owl. I thought to myself that it is an enormous vulture and then as I got closer I saw the white-head of this gigantic bald eagle. It was majestically looking out of the pond across the road to my right. I was in awe. Then a moment later here comes another red-tailed hawk flying now from right to left and flew directly in front of my vision. “Okay, what does this mean?”, I said to myself. I ran back into my office to look up the significance and meaning when it came to native american cultures.

Across many indigenous traditions, the number three signifies completion, balance, and a fully delivered message rather than coincidence. It reflects the alignment of sky, earth and the unseen world, or mind, body and spirit acting in harmony. When events occur in rapid succession, especially involving animal messengers, they are understood as intentional communications rather than random occurrences. I mentioned before being born on the 3rd of January, I have had a connection to the number three and there is spiritual significance of the number across a variety of cultures and religions. It always gets my attention when things happen in threes. Across various tribes the meanings are consistent.

The Lakota (Sioux) tribe believe that the hawk or the Cetan brings messages from Wakhan Thanka known as the Great Mystery. When the hawk crosses your path it means to pay attention to signs and that a teaching is coming. The eagle, which is the most sacred bird, is a symbol of prayer rising to creator, protection and higher truth. Immediately after, seeing a perched eagle watching over you means you are being seen, guided, judged or affirmed. Now the sequence of seeing the Hawk, Eagle and then Hawk within minutes means that a “message comes, creator observes, the message is then repeated with the last hawk.[195]

Through the Cherokee tribal lens, the hawk is known as a protector and a scout so when it crosses your path it is signaling a warning or a directional shift. The eagle embodies spiritual authority so when the eagle appears on the left side it means to to look back at the ancestors, learn from past trials. And the sequence means that “One is moving toward a new direction, but the teachings of the ancestors must guide you.”

The Navajo believe that the hawk represents perspective, farsight and discernment. A

Hawks' instant appearance tells one to shift their awareness immediately. The eagle is a holy messenger who carries prayers to the Holy People. An eagle perched is a guardian that is revealing itself. The sequence of seeing these three birds means that my path is being scanned, evaluated and supported by the Holy People.

The Hopi people see the hawk as a purifier who removes illusions. Its motion from left to right highlights a movement into clarity. The eagle is a Kachina spirit that symbolizes elevation of consciousness! With it watching you, one's vibration is rising or being prepared for a transition. Its sequential meaning is that awareness leads to spiritual elevation and then integration or convergence.

There are plenty more but another would be the Ojibwe who state that the hawk is linked to the east or new beginnings and illumination. When they cross your path, a new insight or a new responsibility is coming. The eagle is the messenger of Gichi-Manido or the Great Spirit and the bird symbolizes truth and the power of vision. The sequence means that “a new responsibility is coming, support from the Spirit is present and wisdom of the past is required.”

It’s a triple sign with a powerful “mirror” effect when looking at numerology coupled with directional symbolism:

Hawk → Eagle → Hawk

Looking at this through the lens of numerology and numbers, three birds are the trilogy, completeness as well as a spiritual confirmation. Many native cultures consider the number three as a completed message, a sign of balance between the earth, sky and the underworld as well as the convergence of mind-body-spirt as one. With directional meanings as the interpretation, the first hawk means "leave the past, move into your purpose." The eagle looking left to right means that the creator sees one's path forward as true and or important with the final hawk representing "Do not forget what you know. A past lesson must be applied now."

Here's where it starts really wild and unmistakable if you were to layer all of these into truth cake and how it aligns with the Green Man of England and Sasquatch of North America. Let's start with Shamanism as the first layer, the next one will blow your mind as it did mine because of what I was writing about specifically. While I was writing this chapter I specifically discussed the idea that I feel "protected" and "directed" on my latest journey. In Shamanic traditions coupled with synchronicity, animal sightings in a rapid sequence aren't random at all, they are synced events and they are often tied to spirit guides, totem guardians or ancestors, nature intelligence and even field consciousness all of which my book is touching upon.

I am traveling along the road, or this journey and the moment that first hawk passed in front of my field of vision, my "vision field" around me lit up. I was now aware of my surroundings, more in the present moment. The eagle watching my path is considered a high guardian signaling that I am aligned, safe and on the correct road or path not only physically but spiritually. An eagle will rarely show themselves without some purpose. Hawk number two was that the message has been reinforced, there is closure and confirmation of my path going forward.

So when you consider Native American traditions, add a dash of numerology, a splash of directional symbolism, then add in a pinch of Old European woodland traditions and the combination of birds, Green Man and stir in some Sasquatch into this cake you get something that falls in the spiritual or otherworldly. The hawk is considered the messenger of the forest and it carries awareness, warnings or signals from unseen intelligences of the land. The eagle is considered a Guardian of the Threshold and symbolizes higher realms watching or approving one's passage.

Birds are often the first to respond when a forest spirit or guardian is aware of someone entering their territory. In Algonquin, Mi'kmaq and Salish traditions, hawks and eagles are companions or scouts of the "wild people of the woods." In Celtic lore, the Green Man communicates through sudden bird appearances, animals crossing your path and directional omens!

In Sasquatch lore, both tribal and modern, hawks will often show up before a sighting or communication and eagles are said to appear in "hot zones" where the veil between the worlds is thinner and the presence of a watcher, in this case the eagle, indicates a boundary area or a liminal zone. I just happened to be driving past Leominster State Forest along route 2 in Fitchburg when this happened! The Land, Monsterland was itself communicating to me. The fact that I am writing about wildman, Green Man and Bigfoot and the connection to nature spirits and synchronicity is astounding to me. It is believed that because I was heading back home to finish writing about this very topic it is as if the meaning is doubled.[196]

It is very rare in many Native American traditions to have a hawk warning followed by an eagle confirmation. For most tribes that combination is essentially "We are guiding you. Keep going." It is considered as a "stacked message" when animals appear in rapid sequence in such a short time period along with directional symbolism that this was an affirmation from the spirit world. Message received. As soon as I arrived home I continued to write and included this analysis of what I had just experienced.

Chapter Fifteen

ELEMENTAL + SPIRIT = INTERDIMENSIONAL

When I speak publicly about the term *interdimensional*, I often describe it as a modern reinterpretation of what earlier cultures called *spiritual*.[197] Across history, humans have understood themselves not merely as physical beings, but as consciousness temporarily embodied—interacting with realities that extend beyond ordinary perception. What we now frame in scientific or speculative language mirrors older descriptions of non-physical intelligence engaging the material world through energy, light, and awareness.

If this is true, then elemental beings, forest spirits, and so-called interdimensional entities may not represent separate categories at all, but different cultural interpretations of the same underlying phenomenon. The Green Man and Sasquatch, in particular, appear repeatedly in environments associated with other anomalous activity such as strange lights, electromagnetic disturbances, missing time, and swift disappearances suggesting not isolated folklore, but recurring interaction points between worlds.

One hypothesis that continues to surface is the portal model: the idea that certain locations act as temporary thresholds where boundaries between realities thin.[198] These areas are frequently marked by geological stress, underground water systems, or heightened electromagnetic activity, and witnesses often describe luminous orbs or flashes of light immediately before or after encounters. Rather than fixed doorways, these portals may behave more like responsive openings appearing briefly, then collapsing allowing entities to enter, exit, or observe.

This pattern becomes especially compelling when examined through well-documented modern case studies. Jacques Vallee said "*the modern UFOnaut and the medieval fairy are identical in almost every respect except for costume and historical setting.*"[199] For example, the forests of the Pacific Northwest, known for Sasquatch sightings, are also hotspots for UFO activity and strange lights. Similarly, the ancient woodlands of England, home to Green Man folklore, have long been associated with mysterious lights and supernatural occurrences. These locations often have a history of paranormal activity, ranging from ghost sightings to unexplained disappearances, further supporting the idea of portals.

This concept has garnered more respect and awareness due to shows like *Skinwalker Ranch* in Utah, which most people are aware is a hotspot for paranormal activity, including sightings of Bigfoot-like creatures, UFOs, and strange orbs or balls of light. Researchers believe that the ranch may be situated on a portal or dimensional hotspot, where entities can cross between worlds. There could even be a craft below waiting to be unearthed. This site provides a compelling case for the interdimensional hypothesis, given the wide range of phenomena reported.[200]

The ranch, named after the Navajo legend of the Skinwalker, a shape-shifting witch, has a long history of strange occurrences. The Shermans, who bought the ranch in the 1990s, reported encounters with large, wolf-like creatures, unexplained animal mutilations, and sightings of strange orbs and craft. Their experiences drew the attention of researchers and the media, leading to a series of investigations.

One notable incident involved the appearance of a large, hairy creature resembling Sasquatch. Witnesses described the creature moving with incredible speed and agility, disappearing into a glowing orb. This sighting, combined with other reports of interdimensional beings and portals, supports the theory that Skinwalker Ranch is a focal point for paranormal activity. In addition to Sasquatch sightings, researchers have documented a range of phenomena, including poltergeist activity, unexplained voices, and temperature changes without warning. These occurrences suggest that the ranch may be a nexus for various types of paranormal phenomena, linked by their interdimensional nature.

Another compelling case from Skinwalker Ranch involves the appearance of a massive, bi-pedal creature observed by multiple witnesses. The creature, estimated to be over nine feet tall, was seen moving through the ranch's fields at night. Witnesses reported that the creature appeared to be enveloped in a shimmering light, suggesting an interdimensional origin. The creature disappeared in an instant, leaving behind large footprints that were subsequently cast and analyzed by researchers.

On March 12, 1977 a large humanoid creature was observed perched up in a tree with unblinking glowing yellow eyes a mere 50 yards away from researchers. It was fired upon and it vanished leaving a single print in the snow. "The Tunnel" incident, described in Chapter 19 of Hunt for the Skinwalker: Science Confronts the Unexplained at a Remote Ranch in Utah (2005) by Colm A. Kelleher and George Knapp, recounts one of the most unusual episodes investigated at Skinwalker Ranch by the National Institute for Discovery Science (NIDS), the research organization founded by Robert Bigelow to examine anomalous phenomena. The event known informally as "The Tunnel" stands out as one of the clearest examples where cryptids and potentially interdimensional elements appear intertwined.[201]

At this point, the recurring pattern becomes difficult to dismiss as coincidence. Across geography, culture, and historical period, similar entities appear in the same environments, interact through the same sensory markers like light, sound, altered perception and vanish in the same ways. Rather than asking whether these beings are physical, spiritual, or psychological, a more useful question may be why the same interfaces repeat, suggesting an underlying mechanism that cultures have described differently but encountered in remarkably consistent ways.

According to the account, two NIDS investigators were conducting a nighttime surveillance operation on the ranch, using high-powered binoculars and night-vision equipment to monitor the property. Skinwalker Ranch, located in Utah's Uintah Basin, was already known for a wide range of high-strangeness reports, including UFO activity, cattle mutilations, and sightings of unusual animals—phenomena often contextualized locally through Ute and Navajo narratives involving shape-shifting entities sometimes referred to as skinwalkers.

During this particular operation, the investigators observed a bright, glowing anomaly close to the ground. They described it as a tubular or tunnel-like structure of light, approximately three feet in diameter. The object appeared to hover in place and emitted a contained, self-luminous glow unlike any natural or technological source. Its behavior and form suggested a structured phenomenon that did not conform to known physical mechanisms.

As they continued observing, the anomaly intensified, and the investigators witnessed a large humanoid figure emerging from within the luminous "tunnel." The entity was described as massive, upright, and roughly comparable in appearance to a Bigfoot or Sasquatch-type creature, with a dark, bulky silhouette. The figure appeared to crawl out of the illuminated opening as though exiting a passageway or portal. Once outside the light, it rose to its full height and moved deliberately across the terrain, at which point the glowing tubular structure diminished and disappeared without a trace.

The investigators interpreted the event as suggestive of a transitory conduit—a temporary opening from which the creature emerged—raising the possibility that the phenomenon represented some type of passage between different locations or even different states of reality. The incident remains one of the most dramatic and enigmatic episodes documented during the NIDS investigation of Skinwalker Ranch.

In addition to electromagnetic anomalies, geological features such as fault lines and mineral deposits have also been linked to paranormal activity. These features may create conditions conducive to the opening of portals or dimensional gateways. The presence of these features in areas with frequent sightings of the Green Man and Sasquatch suggests a potential connection between the physical environment and interdimensional phenomena. This convergence of orbs, lights and UFOs also supports the idea that these entities are not isolated occurrences but part of a larger, interconnected reality.

In England, similar reports have emerged from areas associated with the Green Man. The Rendlesham Forest incident in late December of 1980, often referred to as Britain's Roswell, involved sightings of strange lights and a landed UFO reported by Air Force personnel stationed at RAF Woodbridge and RAF Bentwaters.[202] Lt. Col. Charles Hart recorded his observations with a small audio recorder and later filed a memo of the incident. The craft had strange Egyptian-like hieroglyphics and rumors were that the officers experience time dilation and that the craft was more of a time machine than a space craft.[203]

Interestingly, the forest is also known for its folklore and historical connections to the Green Man, suggesting a possible link between these phenomena! What is being proposed even more is that these orbs are entities amongst themselves and they may be scouts or surveillance systems like Patrick Jackson postulates or even moving portals or openings.[203] We think of portals being fixed and they very well could potentially be opened anywhere or these lights are the doorknobs upon which one can enter.

The Forest of Dean Encounter

In 1962, a forester named William Gardner reported seeing a figure resembling the Green Man in the Forest of Dean. Gardner described the being as having a face covered in leaves and bark-like skin. Following this sighting, the forest became the site of numerous unexplained events, including mysterious lights and sounds.

Gardner's encounter was not an isolated incident. Other foresters and locals reported similar sightings, describing the figure as blending seamlessly with the forest. The Forest of Dean, with its ancient woodlands and rich history, has long been associated with supernatural occurrences, making it a fitting backdrop for sightings of the Green Man.

During the 1970s, several campers reported encounters with the Green Man in the same forest. One group described seeing a figure with glowing green eyes that seemed to emerge from the trees, only to vanish moments later. These sightings were often accompanied by a sense of profound stillness and an overwhelming feeling of being watched.

In 1985, a local historian investigating the Green Man phenomenon experienced a close encounter. While walking through a dense part of the forest, he reported feeling a sudden drop in temperature and hearing an eerie whispering sound. Turning around, he saw a face peering out from the foliage, matching the descriptions of the Green Man. The figure disappeared as quickly as it had appeared, leaving behind a sense of awe and mystery.

In 2020, a family camping in the New Forest, England, reported seeing a Green Man-like figure standing at the edge of their campsite. The figure appeared to be observing them and seemed to blend seamlessly with the surrounding trees. The family described feeling an overwhelming sense of peace and connection to nature during the encounter, suggesting a positive interaction with the Green Man.

The commonalities between the Green Man and Sasquatch extend to their physical descriptions and behaviors. Both are described as humanoid figures with a strong connection to the natural world. Witnesses often report feelings of awe, reverence, and a profound sense of connection to nature during encounters with these beings.

Electromagnetic anomalies have been documented in areas with frequent sightings of the Green Man, Sasquatch and UFOs. These anomalies often coincide with periods of heightened paranormal activity, suggesting a potential link between electromagnetic fields and interdimensional phenomena. Many cultures have their own versions of forest spirits or hairy wild men, suggesting that these entities are part of a universal phenomenon rather than isolated cultural myths.

The wildman is a global archetype. For example, the Yeti of the Himalayas, the Yeren of China, and the Yowie of Australia share many characteristics with Sasquatch, including their physical descriptions and behaviors. As established in the previous chapter, cultures worldwide describe remarkably similar forest beings under different names.

Similarly, the Green Man as well as the Woodwose or Wudeasa have counterparts in other cultures, such as the Leshy of Slavic folklore, a tall forest spirit who can shape-shift, sometimes hairy and humanoid also known as a protector of animals and trees. The Basque County Basajuan is said to be a huge, hairy master of the forest and also considered a guardian of wildlife and the woods.

You also have ones on the smaller side, the Trow of Scotland and Orkney which is a small but hair and wild woodland humanoid reminiscent of the Pukwudgies of the Northeast here in the United States. The Orang Pendek from Sumatra and Indonesia, a short bipedal hairy hominid as well as Japan's Bigfoot the Hibagon.

KNOCKING ON WOOD

"Liminal spaces are where the ordinary rules of reality loosen, allowing other forms of intelligence to appear." - Edith Turner[205]

If portals are not fixed locations but responsive thresholds, then sound, particularly wood-on-wood percussion may function as a method of interaction rather than superstition. Many early cultures believed that trees were inhabited or conscious. The Celtic and Germanic tribes of ancient Europe believed that trees contained Ancestor spirits, Guardian beings or Forest gods.[206] We are coming back to concepts that have been lost.

The early tribes would physically tap or knock on a tree in order to alert or awaken the spirit within. Sound travels through wood in a deep, resonant way and is thought to carry the message straight into the heart of the tree spirit. This was also a preferred communication method to ask for protection. It has survived to the present day as a superstition you might do yourself. People "knock on wood" when they say something that could be interpreted as a curse or negative talk and by knocking on the wood their intention of that action is to negate the thought or words that were just evoked.

So there is a correlation to modern-day wood knocks that bigfoot researchers will often do with hopes of getting a response. This was a great method prior to shows like *Finding Bigfoot* and *Expedition Bigfoot* but now how do you know you are not getting a response back from another human? We also don't know the "code" and how many knocks mean this or that.

Some believe the wood knocks are used as a signal to evoke "this is my territory". What instantly comes to mind is "morse code" as a communication method. Knocking on wood was also a signal to respect the "green ones". In many traditions the act of knocking on wood was described as "calling the spirits of the woods".

Before electricity, forest cultures and indigenous groups also used wood knocking as a way to communicate across vast distances, signaling danger or hunting information. One of my friends is a native american and wildlife guide in Northern California. He gave me a "spirit stick" made out of cedar. It is carved and shaped in the form of the clappers you would find at a college basketball game, small enough that it fits into your hand like a tiny baseball bat and generates the sound of two pieces of wood clapping together. This was the first one he ever made. He has made others that he uses but told me that every time, and he emphasized "every time" he used it in the forest he would bring in spirits, orbs, bigfoot and even dogman would emerge and come to him!

The Salish and Coast tribes believe that sasquatch is known to signal dominance or warn them with powerful tree strikes. Witnesses have also described hearing trees being twisted or pushed over during close proximity sightings or if you are in their territory or hunting ground. The Cherokee tribe describe "Tsui 'Kalu", the forest giant, who uses sticks against trees to send messages through the woods.[207]

When we look at The Green Man, what often appears in carvings and paintings is a branch, a club or staff made of wood which symbolizes a tool of communication. The figure that carries this staff or club commands the forest, understanding the language of the trees and the one who can make the woods or forest "answer back".

The Medieval Woodwose or wildman is known to make booming knocks in the forest or drumming sounds, tree knocking during mating or territorial displays. The Basque forest giant, a spirit helper, pounds on trees to warn shepherds of danger. This also aligns with “the Watcher of the Woods” which many native american traditions believe that the “wild man” spirit carries with him. The Mi’kmaq share that their “wildman of the woods” uses wood strikes as well as stone knocks to communicate.[204]

The Green Man, as an elemental, embodies the spirit of nature itself, representing growth, renewal, and the cyclical nature of life. His appearances often coincide with significant environmental events, suggesting a role as a protector and preserver of the natural world. Sasquatch, similarly, is seen by many Indigenous cultures as a spiritual guardian. The frequent association of Sasquatch sightings with natural disasters or environmental changes supports the idea that these beings are closely tied to the health and balance of the natural world.

The Interdimensional hypothesis offers an explanation for the elusive nature of these entities and their ability to appear and disappear at will. If the Green Man and Sasquatch are indeed interdimensional beings, their sightings would naturally coincide with other paranormal phenomena, such as orbs and UFOs, which are often reported in the same areas where there is electromagnetic and paranormal activity.

SASQUATCH AS A SPIRIT

I was part of an event not too long ago that incorporated Ghost Hunting, Bigfoot and UFOs, called the Trifecta. An idea that Rob, Sarah and I from Get Haunted had come up with and their team executed it into reality. The idea was to incorporate the three into one event as there may be a connection. We had an amazing experience at the Haunted Shanley Hotel and the team located a forest to investigate in New York for Bigfoot. We also spent time at the Pine Bush Museum for the UFO portion of the event.

Bill Bryan of Empty Casket Paranormal entered the event firmly grounded in the flesh-and-blood interpretation of Bigfoot, approaching the phenomenon from an investigative and engineering-driven mindset. He was curious but never had ventured into that realm of thinking and investigating. So this was his first time doing something like this. He created a woodknocker and gave me the first prototype which we used while we were out in this forest. We did a series of knocks and apparently something responded. I asked him to write up his experience to share in the book because it is wild!.

John E. Mack said that "the experiences appear to occur in a realm that is neither purely physical nor purely psychological, but somewhere in between."[208] Bill recalls his experience. "I have always been firmly in the flesh-and-blood side of the Bigfoot debate. But I have had a few experiences that have left me scratching my head and opened me up to the possibility of there being more to Bigfoot."

"The first being at the 2024 Get Haunted Trifecta event at the Shanley Hotel in Napanoch, NY, a paranormal event that combined ghost hunting, UFOs, and Bigfoot. During the bigfoot hike with Ronny Leblanc, there was one field that caught my attention; I badly wanted to go into it. But we continued on further down the trail. We received a direct wood-knock response, followed by an EVP instructing us to leave the area.

Bryan continues, "the part that really got to me was on the way back out of the woods as we passed that field again. Once again, I was drawn to it. As we walked past, my eyes were glued to the field. What I saw resembled the cloaking effect from the film *Predato*r—a distortion where the background remained visible, but the outline of something solid was unmistakably present.

Branches were hanging down, there was tall grass, and it was a hot day. So, I figured that it was just the branches and grass interfering with each other, causing an illusion. If that is the case, then I should be able to recreate it. So, I took several steps back and then forward again to see if I would see it again. I didn't, so I tried it a few more times."

"At this point, everyone had stopped and was watching me. After several attempts to see it again, I yelled, 'What the hell was that!?' to which I heard the best thing I could have heard at the moment a woman behind me said, "You saw that too?" I knew this was not a solitary misperception.

Ronny immediately asked me to describe what I saw. After telling him, he said I had seen a glimmer, a bigfoot cloaking itself. *My mind was blown.* Did I really see a cloaking bigfoot? That would throw my whole flesh and blood only thoughts on bigfoot out of the water. This opened me up to the possibility that bigfoot was more than just an undescribed relic hominoid. But as time passed and I reflected on it more, I concluded that I don't know what I saw. It could have been a cloaking bigfoot, or an alien, a spirit, or one of many other possible things, either spiritual or mundane. The problem is that I just didn't have enough information to know."

I saw Bryan at another event in Maine this past year and he was freaking out to share what has happened since that time in New York.

"Well, it wasn't until the 2025 Get Haunted Trifecta event, this time at Fort Duffield in West Point, Kentucky that it happened again. This entire event was outdoors in the woods. As dusk settled in, my friend Kayleigh came looking for me to show me some of the pictures she took behind the fort. Among them were pictures of what appeared to be a bigfoot (Figure 1) and a Grey (Figure 2). So, I asked that she show me where she took those photos.

We went behind the fort to look. By this time, it was starting to get dark, but not so dark that you couldn't see. We started seeing shadows walking around the fort. Whether they were bigfoots or spirits, we didn't know. Then Kayleigh pointed to a small hill and said that is where she took the photo of the bigfoot.

As I crested the hill, only a few steps away I nearly walked directly into it. Standing on the crest of the hill, there it was. I was watching in front of me.

There was nothing there...and then there *was*.

There was an outline, a very clear outline, of a bigfoot! In the middle, I could still see the forest behind it, but around the edges was that same glimmer-like effect (Figure 3).

Walking up, it wasn't there, then suddenly it was, less than two feet in front of me. Just as quickly as it was there, it wasn't. I screamed and jumped back off the hill; it was gone. I apologized to it for screaming and almost walking into it. Kayleigh and a few others came running over, having heard my scream. I told them what I saw, and they went to see for themselves, but saw nothing.

The bottom line is this fully upended my world. The rest of the night, I told everyone what I saw, hoping someone could give me any explanation other than it was a cloaked bigfoot. But that is what everyone told me, it was a cloaked bigfoot. The drive back to the Airbnb was full of thoughts on what I had witnessed. This experience is all that I thought about on the long drive back to Pennsylvania. What did I see? What was it? Is bigfoot really more than just flesh and blood? Was that the same kind of thing that I saw a year earlier in New York with Ronny?

As I write this, seven months after seeing it, I am still questioning 'What did I see that night in Kentucky?' It has upended my view on bigfoot and what it is. There are stories about the soldiers stationed at Fort Duffield during the Civil War, shooting at nothing during the night. What did they see? What frightened them? Did they see the same thing that I saw that night? If so, I can understand why they would be frightened to the point of firing their guns, seemingly, at nothing."

Bill's encounter dismantled the boundaries he had placed around what Bigfoot could be. He could no longer dismiss reports of cloaking, disappearance, or non-physical behavior as an exaggeration. What he saw may have been a biological entity, a spirit, or something that occupies the space between. Now Bill can no longer say it was impossible.

Bill has a theory of what he may have seen, what others may have seen and potentially are seeing. We talk about the *interdimensionality* of these beings and how that fits squarely into the spirit bucket, but it also doesn't explain the visual sightings, rocks being thrown or structures being discovered, but it makes sense what he postulates. If humans can leave impressions, apparitions, or residual intelligence after death, then it is reasonable to ask whether other intelligent beings might do the same?

"There is only one explanation that I can come up with that does not remove me from the flesh and blood camp...it was a ghost of a bigfoot. Bigfoot is thought to be close to humans, anthropologically. So, if humans can become ghosts after death, why not bigfoot? If you take the bigfoot stories, such as knocking, rock throwing, unexplained smells and noises, footprints that come and go without explanation, sightings that appear and disappear out of nowhere, etc., and take them out of the forest and put them into a house, you are describing a haunting. So, then is bigfoot just a flesh-and-blood creature, but the otherworldly aspects of bigfoot stories actually encounters with a bigfoot ghost?

The truth is that I simply don't know. I don't know what I witnessed. I don't know what it means to do bigfoot research. What I do know is that I can no longer brush off the more paranormal bigfoot sightings and experiences that witnesses report. I know that I now have a more open mind to the possibilities of what bigfoot may or may not be. And I know that I will continue to seek answers. The truth is out there."

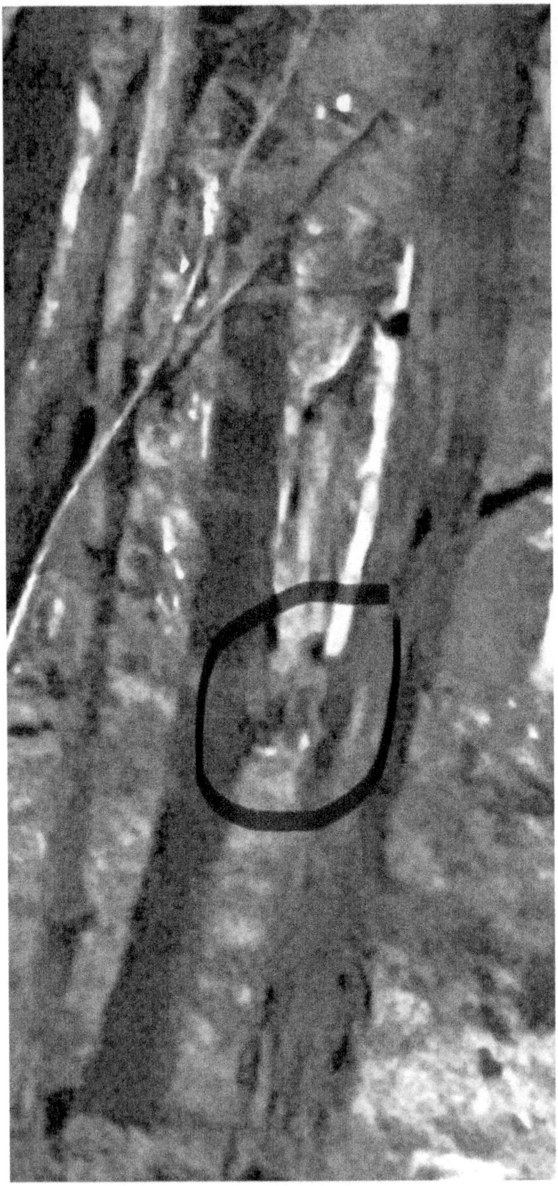

(Figure 15.1a Possible image of a partially visible humanoid form consistent with reported Sasquatch morphology, captured at Fort Duffield, West Point, Kentucky during the 2025 Get Haunted Trifecta event. Photograph by Kaleigh Baysinger, 2025)

(Figure 15.1b Cropped enlargement of Figure 15.1a, highlighting a specific area of the image where the humanoid outline appears more clearly. This detailed view is provided to emphasize form and proportion rather than to suggest a separate or additional sighting. Photograph by Kaleigh Baysinger 2025)

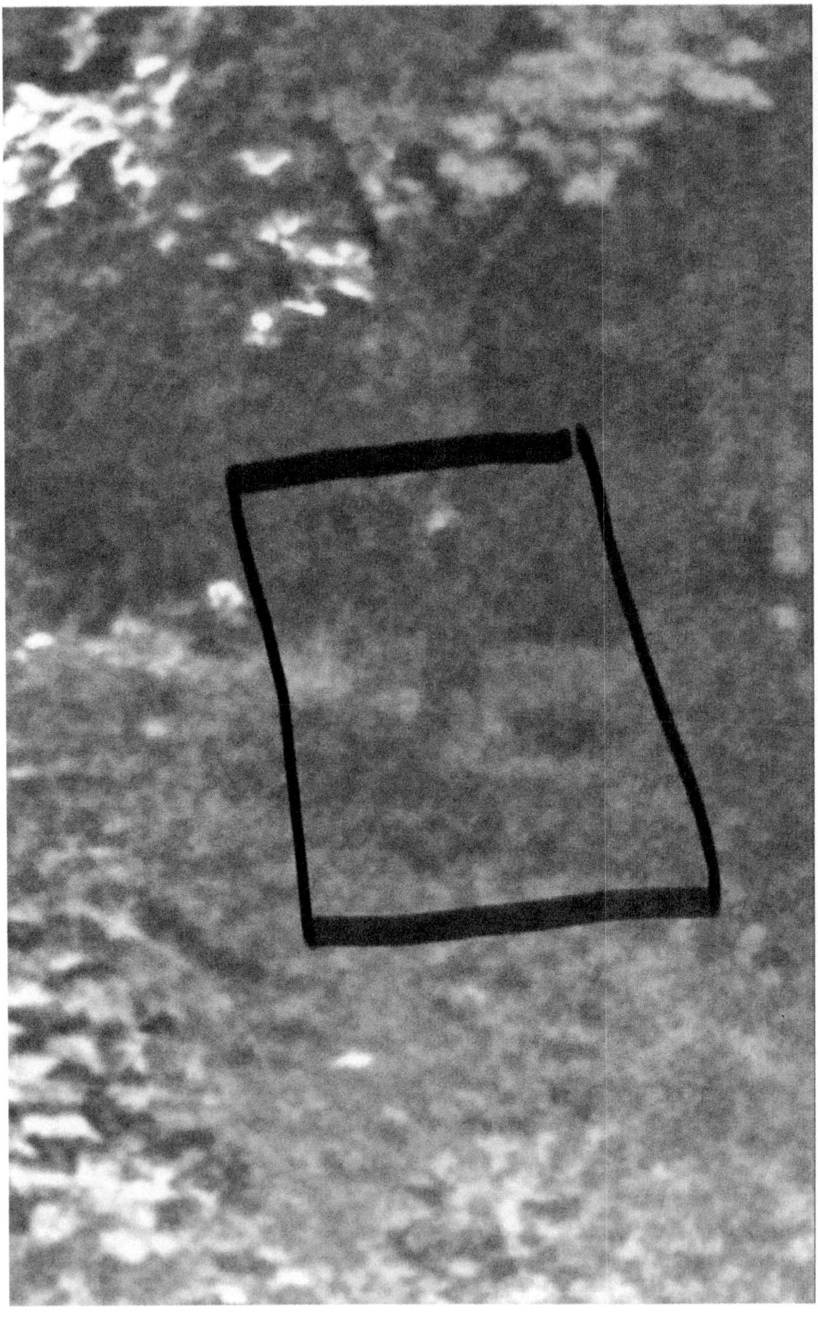

(Figure 15.2 Photograph taken near Fort Duffield, West Point, Kentucky, showing an anomalous distortion or absence consistent with witness reports of transient visibility during close proximity encounters. Photograph by Kaleigh Baysinger 2025)

(Fig. 15.2 Artistic rendering made by Bill Bryan depicting the cloaked humanoid figure he observed at Fort Duffield, in West Point, Kentucky 2025, based on direct eyewitness experience.

Chapter Sixteen

BLOODLINES OF THE VEIL

What I have come to recognize after years of listening to experiencers is that the phenomenon rarely moves through individuals alone—it moves through families, landscapes, and bloodlines. When encounters begin to cluster across siblings, parents, children, and even grandchildren, especially within the same geographic corridors, the randomness disappears and a pattern takes shape. Chapter Sixteen is not simply a collection of extraordinary sightings, but a case study in how awareness, sensitivity, and contact seem to propagate through kinship itself.[212]

What follows are two parallel family narratives, one unfolding across Indiana and Washington, the other rooted in Kentucky. They both mirror one another in timing, escalation, and consequence. Together, they suggest that when the veil thins for one member of a family, it rarely stops there.[213]

SUZANNE FROM INDIANA

I have always been talking about synchronicity being an awareness tool when it comes to interacting with the phenomena or interpreting people who come into your life as confirmed by acts of synchronicity.[214] I used to be aware of this and not necessarily take action on it. I have talked about when it comes to writing my books that I feel led in certain directions and for whatever reason people seem to find me as opposed to the other way around.

Back in 2023, at the Smoky Mountain Bigfoot conference in Townsend, Tennessee, I was one of the speakers and was sitting at my table getting ready for the crowd as the doors opened on the first day of the event. My table was directly near the entrance. There was a recent and compelling video making the rounds on social media of what looked like a humanoid-type translucent figure being filmed in the forest by a couple going for a hike.

You could see its definitive shape, and like the movie *Predator*, it was trying to remain invisible, crouching down and stopping then moving. I got the energy from the video that it was authentic. You could hear the voice of a woman wanting to get the heck out of there. It didn't seem to be manipulated in any way. I was watching it on my phone as it lay down on the table in front of me when a couple, one of the first people to enter the conference, approached me.

The younger woman and her brother approached me smiling and when she looked down at the video playing on my phone she had a visible reaction of shock. She looked at her brother and looked back at the video. "What the hell is that?!" I picked up the phone and started the video over again and came around so they could see and we watched together. "It is a bigfoot cloaking or some sort of cryptid. I think it's real", I stated.[215]

"Okay this is so weird. My Mom just told us yesterday about an experience she had with something invisible just like this!" Synchronicity bells started ringing. "No way! Would she be able to chat about it?" And at that moment, without any hesitation she picked up the phone and dialed her mother. "Mom, you are not going to believe this. She explained everything to her Mom and handed me the phone. I exchanged my contact information with her so that we could do a proper interview in the next few days. What are the odds that their mother tells them this experience out of the blue and then a few days later they come to the conference and see me watching a video of a similar phenomena. Suzanne resides in Nineveh, Indiana.

"Hi Suzanne, thanks for taking the time to speak with me. I am sure that this was the last thing that you were expecting", I laughed. She giggled back and in a sweet grandmaesque tone that put you at ease, "I am happy to talk to you. I didn't realize that others are seeing similar things out in the woods."

"Now, I don't know where you want to start but what I saw happened this July. I was sitting on our upper deck at our new house with a view of the woods, the deck is about ten feet off the ground, there is a ridge behind us and power lines that run through the back. We are surrounded by trees. I don't remember that exact day, but it was around 1 or 2 in the afternoon and it was a bright sunny day. I was just sitting out here and looking out and immediately I saw this oblong-shaped burry object. It looked as if it had energy waving in a clear shape. I didn't see an outline or anything, it was just a shape. It was in a vertical position but it was way high up in these trees."

"I noticed it on this cluster of leaves and asked myself 'what is that?' and this object, it just started slowly moving across from one tree to the next, and then to the next and it probably covered a span of about six or seven trees."

"Was it moving from tree to tree, like it was in the trees?' I questioned. "No, it was more like it was gliding from tree to tree, like floating.[216] I could still see the foliage through it. It looked like there was energy moving through it as it moved across. I started to get an eerie feeling when I was watching it. It finally dropped down near my husband's shop and then disappeared. I just thought to myself what was that and the fact that I hadn't seen anything like that before so I just kept it in my memory."

As Suzanne described it, the object did not move like anything biological. It glided—quietly, deliberately—passing from tree to tree while remaining partially visible, as though the forest itself were momentarily out of phase. What unsettled her most was not fear, but recognition: the sense that what she was seeing was aware of her presence.[217]

"About a couple weeks later my sister and my Mom came to visit me. We were sitting out on the porch and I told her about that strange experience I had while I was sitting out here staring out into the woods. I asked her if she has ever seen anything like that at her property. Their house is in the middle of the forest as well. She lives about two and half hours away. I asked her if she had seen anything similar where she lives and she said 'Yes, I have'. I was stunned."

Her sister then described her experience which was just a few weeks before Suzanne's! She said that she was in a similar state just sitting out by the pool and enjoying nature when she said she went inside to grab a drink and came out and her entire yard instantly looked blurry. It was the whole line of the woods and it stayed that way for about two to three minutes and then it disappeared. It was misty looking and covered the entire tree line."[218]

As Suzanne was describing this to me, her experience and her sisters seemed to be similar but different. Suzanne was seeing a figure of sorts moving and her sister, almost as if a wall or veil was revealing itself, like a portal. Either way, something neither of them have experienced before. What was intriguing is that they both happened within two weeks apart and a driving distance of about two hours from one another. They also both happened around the same time, around 1 pm or 2 pm in the afternoon.[219]

Suzanne then describes that her grandson, while he was visiting, witnessed something at their property. He's 14 and lives in Olympia, Washington. He was willing to write something up describing what he saw at his grandmother's house. Suzanne asked them all while her kids were visiting for a week that if they saw anything strange or weird at her house to let her know. She specifically described what she had seen so they knew what to look for. Her grandson ended up telling his mother when they got home what he saw as he was a little embarrassed to say anything while they were in Indiana staying with the rest of the family.

"I was visiting family in Brown County, southern Indiana and I was staying at my grandparents house. Their house is in a forested area and it is on a hillside. So the backside of the house has a balcony that is about ten feet off the ground and overlooks the forest. It was daytime and I was sitting on the balcony with my family. I looked over at the forest and saw a blurry mass and it almost looked like what my grandmother had previously described. It was hovering in the air between the balcony and the start of the forest.

The distance from the balcony to the trees is about 15 to 20 feet. So I could tell that the mass was the size of a bowling ball. It was only visible for a few seconds but it slowly changed its shape. It moved behind the railing of the balcony and disappeared.[220] I also have never had any problems with my vision. I have a theory that it could be some kind of ball of gas that was floating in the air but I really don't have any idea of what it really was."

Suzanne was inside when her grandson witnessed this bowling size transparent mass. "Now Suzanne, have you had any experience before this or leading up to these experiences because they are all around the same time period, which is about six weeks?" She paused for a moment and said yes, she did. I wasn't surprised. What tends to happen is an uncoiling of strange occurrences that span the experiencer's lifetime, their siblings as well as their parents and other family members seem to have similar encounters that they often don't even share with one another.

"I did see something though and when I did see it, I didn't know how to categorize it. So, I kind of put it into a 'spirit folder' I guess." This occurred a few weeks before all the other events. She and her husband take a particular road that goes through the forest leading to the interstate. They take it all the time but she had hit a deer on this road so any time that they drove it she was on high alert when her husband was driving. They were on their way home and this was about three or four o'clock in the afternoon and the sun was still high in the sky. The road curves to the right and left and there is a small section where there is a straightaway where you can see ahead of you. There are woods to the right and a field on the left of this stretch of road. The field on the left butted up to a heavily wooded ridge.

As they rounded the bend just yards from their driveway, Sherry glanced down briefly, catching a flash of red in her peripheral vision. At first, she dismissed it as early autumn foliage—until the realization hit that the leaves had not yet begun to change. When she looked up, time seemed to slow.

Standing at the edge of the woods was a figure so large and unmistakable that disbelief lasted only a few seconds. The auburn hair caught the sunlight with an almost polished sheen, framing a broad, powerful form that was both ancient and alert. Its eyes were a bluish-green color and the eyes were deeply set beneath a pronounced brow locked onto hers, and in that moment Sherry sensed not threat, but mutual surprise.

The encounter lasted only seconds, yet every detail imprinted itself with unnatural clarity: the immense shoulders, the hairless gray skin of the face and hands, the arms swaying slowly as if suspended in another rhythm of time. It did not advance. It did not retreat. It simply stood—aware, present, and watching, before the moment collapsed and the forest reclaimed itself.

As we approached the straight and narrow I was looking straight ahead to make sure there were no deer about to cross the road. This was crazy, I have never seen anything like this before either, but something bolted out of the woods like it was already in a full run and ran across the road so fast and disappeared as it got into the field and I couldn't see it any longer. It wasn't a solid animal. It was transparent and it had a greenish or grey tint to it so I was able to see it. I could see legs in motion as it was running![222]

I turned to my husband and asked "What the hell was that that just crossed the road?!"

"He hadn't noticed it. He didn't see it as he normally is in his head, not really paying attention to his surroundings when we are driving." Suzanne states emphatically that it was not an animal. "It wasn't solid. It wasn't flesh and blood. It looked as if it had two legs but it was humped over and kind of had a transparent look to it. When I looked over to the field, I couldn't see anything. It just vanished. It looked larger than a human being and I could see through it. I could see the motion of it when it was running. It was a clear day. I thought that it might be a spirit or something, but a spirit of what though? I didn't look like a human spirit to me."

How far away was this sighting from your house? "About a five to ten minute ride to that section of the road." When did you move here? "We just moved in only a few months ago. My husband has been building this house for over a year and a half. I will also say this, I don't know if my kids told you or not that I seem to be connected to the spirit realm or sensitive to it. This was an energy I wasn't familiar with. I can tell you when it comes to ghosts or spirits whether it is a feminine energy or a male energy. I consider myself a clairvoyant. This energy did not feel human. There was a wispyness that came off of it as it ran. Like wisps of energy. It was moving so fast."[221]

I asked Suzanne if her parents had this ability to see things that others can't see like ghosts and spirits. Are they sensitive? She said she found out that her Dad who is now 87 just saw a ghost of a little girl at his house. He admitted to her that he has had this ability since he was a child yet he never talked about it.[223] She said that she has had the same ability as a kid but never discussed it with anyone until much later in life. It is unfortunate that when we have these "gifts" there is a stigma around talking about them, sharing these experiences or sightings even with members of our immediate family. She went on to tell me that her Mom has premonitions and can see things before they happen.

As I continued on with Suzanne's interview my questions continued with confirmations and boxes being checked when it came to her experiences. Now I was confident that she also had some UFO experiences as well. This could be how her parents may have received their "gifts" and she as well. Experiencers often receive psychic capabilities and other gifts after UFO sightings, abductions and or encounters with other beings.

I dove right in. Have you had any UFO experiences as a kid? "Well, this is going to sound crazy and I never told anyone until I told my other grandson, the older brother of the one that saw the anomaly in my backyard. He was with his other brother and they witnessed a triangle UFO following them and hovering over the tree line in Olympia, Washington. They ran home crying and terrified. It was large enough that it blocked out the stars and all they could see was this inky black triangle hovering about them.[224]

"The oldest brother presumes that he has had more UFO incidents than anyone else in the family. He shared the story with me so I wanted to share my experience with him when I was young. And again, this is going to sound crazy and I never told anyone this until he came to me." I quickly reminded her that she was preaching to the choir and I had experienced very similar things so nothing shocked me anymore.

"I don't want you to think I am a cook, crazy, but this was bizarre. This happened back in the 1990s in Van Buren, Indiana. We lived in this big old Victorian style house. It was really early in the morning but it was still dark out. He was up and went out of the room but had left the light on in the closet bedroom. I opened my eyes up...now I have never seen a UFO before, but when I opened my eyes I saw three grey aliens standing there and just staring at me. Two in the back and one in the front. They were hovering. Their feet were not touching the ground. They were just staring at me over by the window. I blinked. I rubbed my eyes but they were still there."[225]

"I told myself, Oh no way. I haven't seen aliens before. No, no, you are not there and as I watched then the two in the back just glided backwards through the wall and they were gone. The first one was still staring at me and then just glided through the wall behind them after a beat and vanished. I told myself that I would never tell anyone this story. It was something that I never thought I would experience in my lifetime. I still haven't even told my husband about it."

This was her only "alien" experience...for now. I explained to Suzanne that the greys have been reported to show up in threes when they visit.

What unsettled me most was not the number of experiences Suzanne described, but the way they escalated once awareness set in. Curiosity had given way to attention, and attention appeared to invite response. I had seen this progression before, and I would see it again soon, in Kentucky, where another witness would knowingly follow the same path with consequences that couldn't be ignored.

Her sister, 61 years old, also witnessed a UFO where she lives which was a flying saucer or spherical shaped craft that first was a light then went dark. It started as a beam of light shooting across the sky, flying over her head and then it just stopped over the trees. It was the size of a small airplane but was round. It was a mere few feet away. She saw this with several witnesses including her husband. It was hovering over the trees and it was so close that she could see what resembled a big windshield and beings or figures were moving inside the craft! They were doing something and she was convinced that they were looking at her. It hovered motionless and then it was gone. I sensed a connection.

It gets even weirder. I asked if she had any other UFO sightings prior to this. She had another sighting near the local reservoir with two other people, her brother and his girlfriend when she was about 15 or 16 years old. It was the size of a car and was coming out of the sky and heading in their direction right for them. They were frightened and freaked out and left. Two years later someone came forward on social media stating that others had witnessed a similar sized craft over that reservoir as well. The reservoir is about fifteen miles from her house.

During the interview, as Suzanne's sister was about to change gears and share a bigfoot encounter while she was driving, we heard her say "uh oh" and even though her connection was still there we couldn't hear her anymore. It was strange. Suzanne and I continued to talk and she said something strange had happened recently that she wanted to share with me. She happened to just come back from her sister's house on Sunday and when they arrived back to their house, she was sitting on the couch and it was dark outside.

She noticed some strange white opaque thing, the size of a salad plate, oval-shaped and was outside in the top part of her window of her living room that faced back towards the woods. It had tendrils and looked more like a jellyfish. We are hearing more about these types of UAPs more frequently now.[226] This object was directly opposite of the window and then squeezed itself at the bottom and the tendrils moved and it propelled itself up and out away from the window and into the darkness into the forest.

She then remarked that this property fell into their laps. They weren't really looking for it. "For a while, whenever we would get to the property and I would start walking around I would get really dizzy and light headed. I wondered if I could even live here because of the way it was making me feel. The energy was intense and strong. I don't know if it is the powerlines down the way or what but I have somewhat grown accustomed to it."

When Suzanne was living at their old house, she had her right foot up on the couch and her ankle was exposed and she noticed a strange brown triangle shape under her skin.[227] Her entire family including her husband noticed it right away and began looking at it to see what it was. "Where the heck did that come from?" asked her husband. It was disturbing her and she said she grabbed a needle to start picking at it. She wasn't sure if it was a rock embedded in her skin but as she began to touch it - it then vanished the next day and she never saw it again! It sounds a lot like either a tracking device or evidence that she is being abducted. There may be others in her family that are also being taken.

Her oldest grandson that was living in Olympia, WA now lives in Indiana. Ever since he was a little kid, he has memories of being on UFOs. When he was really young he would tell her about a light coming into his room and taking him. He describes being inside their ship. They lived in Okinawa, Japan for three years. While living on the base, Suzanne's daughter said that she was talking with her son when he was in his bedroom. He was sitting there with his laptop on the bed. She started vacuuming throughout the house and when she came back to his room he was totally gone.

"He was gone. I went through the entire house and he just disappeared! I couldn't find him anywhere."

She came back to his room again after frantically searching and there he was back again. "Where did you go!? I was looking all over for you!" He responded that he was here the whole time. He never left the room. "No, you were gone. You vanished." This completely freaked her out based on what he has been telling his Mom for years since he was kid. This also happened to Betty Andreasson when she was little.[229] She disappeared during a day family party and was gone for hours and then reappeared. She sensed she was only gone for ten minutes. People are not just leaving their body when they are sleeping and dreaming. People are completely disappearing and going somewhere and then returning.[230]

When her grandson traveled back to Olympia, Washington to visit his other grandparents he kept setting off the metal detectors at the airport. The TSA pulled him out of the line. They checked him and when they used the wand there was no issue on hits until they scanned his right ankle. The wand quickly went off. It sounds as if Suzanne and her grandson and maybe other members of the family have some sort of alien implant. When he got to Olympia he and his Mom used a magnet and it stuck directly to his ankle![228]

He said his first experience he remembers happened in Fayetteville, Indiana on a military base during a time period when there were UFO sightings in that area according to a MUFON researcher who also witnessed a UFO in the immediate area around this same time period and found out later that he was actually abducted. The UFO researcher's buddy confirmed that he was gone longer than he thought he was!

Suzanne also revealed to me that her son-in-law, the father of her grandson with all of the UFO experiences, is the leader or captain of a Special Forces team. He has spent time in Afghanistan on several tours. There's the military connection! It gets weirder. Suzanne was crocheting in her living room and started to hear noises coming from the forest. She was compelled to go and investigate. So she started to film and did about five different videos. In one of the videos she says what looks like a face that had black eyes and sharp ears...but the being was green and matched the foliage. When Suzanne showed her son-in-law the video and he saw the creature or being immediately. Almost as if he had seen it before.

In the video she sees something moving and it looks bipedal but with that cloaking effect. You can see through it. She describes seeing what looks like head and shoulders. The shoulders and the head are bouncing up and down like something or someone that these are connected to is walking. Something to note is the fact that the blur she witnessed not far from her home while driving with her husband was also green or grayish in color. Could it be the same being?

The grandson who has had UFO encounters now lives with his girlfriend. His girlfriend never had any type of alien or UFO experience until she lived with him. She has shared with Suzanne that she has had several "dreams" seeing large tall aliens at the foot of her bed. They live in Indiana.

Suzanne believes that when she was young she did something that would be equated to teleportation or astral projection. When she was little, 7 years old, she was going to see her cousin the next day and she was so excited and while she lay in bed she was trying to figure out a way to see her cousin now - she didn't want to wait. She then found herself out on the street in the dark in what could be described as a vacuum state. She didn't have to breathe, there were houses and lights but no people. She could feel herself walking to her house and there was a thing, a monster that had white glowing eyes and started to follow her home.

It was hairy, huge with what looked like no neck. She told herself that she needed to get home and wake herself up to get out of this mess. She thought to herself that what she had was a nightmare. She still remembers it so vividly as if it was a real memory. She saw the same thing when she was thirteen. She was sitting on the porch with her grandfather in Kentucky. The creature was across the street and crouched down near a bush. The eyes were glowing white. She thought that it was a bear at first.

The creature then stood up, puffed out his chest, the eyes still growing. Her grandfather couldn't see it. Her aunt then arrived home from a date and the car headlights illuminated the bush but there was nothing there - it just disappeared. She also had another experience with this being in Marion, Indiana where she grew up as well.

Three years ago she was visited by a spirit and this being was following behind the spirit. She was not only seeing it with her mind's eye - it had found itself inside her home sitting in the corner looking like a dark shadow. She described it as looking just like a bigfoot. Its energy was so strong and it was scaring her granddaughter's cat. When she lived in Tennessee she laid down and as soon as she closed her eyes. She had a vision immediately of the sky opening up down the road from them and there was a yellow light and the bigfoot creature emerged from the light! She could see it walking up the road towards their house. It was similar to a movie playing and as clear as a real experience.

She doesn't understand why she sees these things and why she is having these experiences. When she sees spirits they look as if they are solid, living people. Her husband has also heard a growl and snort coming out of the thin air in their bedroom when she had this experience. She heard it as well but at a separate time when she came to bed. She has kept these experiences and encounters to herself and her family. I was privileged and grateful that she was comfortable enough to share them with me.

About seven years ago, her sister Danelle and her neighbor both had fires going one night and they heard a growl that scared her to death. It moved between the two fires but they were unable to hear any footfall, leaves or branches or see what was generating the noise. One of Danelle's friends was at the campfire with her daughter, a little girl who was so terrified by what was going on they had to leave as the growls continued between the two fires. Her neighbor and her neighbor's fiance were the only ones at the other campfire. Danelle was now all alone. She yelled out to Danelle, "Can you hear that!?" Danelle responded quickly "Yes!" "I am coming over to get you with a flashlight," her neighbor called out. Danelle responded, "Please do!"

She came and got Danelle and when they got back down to her neighbors campsite the thing that was making the noise was right there in the woods screaming at the top of its lungs. "It was awful and sounded like nothing I have ever heard before. It was bone-chillingly scary. The part about not hearing it walking back and forth really scared me," said Danelle.

Both sisters see strange cloaking creatures, UFOs, Bigfoot and spirits.

At this point in my research, Suzanne's story no longer perceived to be isolated. The clustering of cloaked figures, UFO sightings, Bigfoot encounters, generational sensitivity, and geographic relocation formed a familiar rhythm—one I had begun to recognize elsewhere. Not long after speaking with Suzanne, I would encounter another woman, raised only a few hours south in Kentucky, whose experiences followed the same unsettling progression: curiosity, invitation, escalation, and consequence.

What struck me most was not simply the similarity of what was seen, but how it unfolded—through prayer, belief, fear, and ultimately confrontation. Like Suzanne and her sister, Sherry's experiences would reveal that when something answers the call, it does not always arrive in the way we expect.

KENTUCKY SHERRY

Sherry Palardy was born in Madison, Indiana, but the experiences that would ultimately change her life took place in Kentucky—along the same Appalachian corridor that has quietly threaded its way through so many of these accounts. I met Sherry after a conference presentation in Kentucky, where she approached me with hesitation, the way people do when they are unsure whether speaking out will bring relief or consequence. What she shared would echo the same progression I had already heard from Suzanne's family: belief, invitation, escalation, and encounter.

Back in 2020, Sherry and her husband Robert became interested in watching anything on television that had to do with Bigfoot. Almost fanatical about it that they didn't want to miss any episode of a television series or any documentary that got released. It became a focal point of Sherry's that she now was so fascinated by all the stories, sightings and encounters she really wanted to have one for herself. Despite being skeptical, she was like many people hoping to have their own personal encounter, although the thought of it being scary, a personal encounter would be the definitive way for her to determine their existence.

"We loved our road trips, country views, and Robert and I would always talk about 'I wonder if we would see Bigfoot today' became a common hope during our adventures. I got to the point I would yearn so badly to want to see Bigfoot. Being a strong believer in Jesus Christ, my upbringing was an old fashioned Pentecostal background."

This was something that I could relate to myself being brought up in an evangelical church in Massachusetts, if it wasn't of God then it had to be of the devil. A sense of fear to go down the wrong path in life or not following things like the Ten Commandments would put your soul in jeopardy. Being brought up this way, we always were told to speak the truth and not lie about anything. Sherry stated that making up stories or lying about something was just not a part of her life or how she was raised. Like most Christians, God was always watching.

"Well hardship began to affect me and my husband's life. Robert's father became ill back in Connecticut so we decided to take the opportunity to help him out and moved to Connecticut briefly in 2021 to care for him, eventually returning home to Kentucky a year later. We took our time heading back and focused on the road trips we enjoyed so much. This ignited that urge again to see a bigfoot. It got to the point that Sherry would pray out loud "Lord Jesus...if Bigfoot is real, then I want to see one. I want to be kept safe, but I really want to see one!"[231]

This instantly reminded me of Bill and Julie Penning the day they discovered a series of six bigfoot tracks in Leominster State Forest back in 2010. Bill said that he woke up that morning and got out of bed, got down on his knees and prayed to God that he would give something for him and his wife Julie to connect with. Finding those prints and feeling like they were being watched as they were alone in the forest brought me into this world of weirdness too as I casted one of the prints and eventually my first book Monsterland was born from it.[232]

"It got to the point on our road trips that I wanted to see Bigfoot so badly that I would actually pray and ask God to intervene. Lord Jesus...if Bigfoot is real I want to see one, I want to be kept safe but I do want to see one." When Sherry returned to Kentucky they were staying with friends looking for a place to live. She had totally forgotten about this particular prayer with hopes of having her own Bigfoot experience. She was also praying that God would provide a place to finally call home.

"About a month later my prayer was answered. We found a home to rent and it was actually more than we had expected. God opened a door for a beautiful lake front home on Lake Cumberland with a beautiful river nearby in Nancy, Kentucky. A gorgeous property nestled next to Daniel Boone Forest. I didn't think of the prayer I requested with so much emotion to see a bigfoot as we were both so excited to finally move into our new home in December of 2022."

"This beautiful home, which was two floors, had a porch downstairs and an upstairs balcony coming out from the master bedroom. So the very evening we were all moved in I decided to step out onto the upstairs balcony seeing the beautiful view from this height for the very first time. As I was standing there all alone admiring the scenery and our new backyard I began to hear this grunting sound. It was like the snorting of a hog and it was so loud! It was as if the sound was amplified 100 times louder than any hog I had ever heard before. Being a country girl, having been on different farms with hogs, it was unlike anything I have ever heard before."[233]

Sherry continues to describe the shift from admiration of the property to instant fear. "Still not even thinking bigfoot, the evening intensified into a night that I sensed we might be in danger. I didn't know what was in the woods behind our home, but whatever it was it sounded very angry. The hair all over my body begins to stand straight up. I was in shock just listening and then all at once I heard a big tree being slowly twisted. It cracked and popped and then fell to the ground. And when it went down, it went down with force! At that point, after being frozen in one spot hearing all of this I ran inside and began to make sure every door and window was locked! I loved our beautiful home but at that point I began wondering...are we safe? What have we got ourselves into moving here? What is in those woods behind our home? That's all I could think about. Still, Bigfoot never crossed my mind at that point."

"When I went back inside after putting our home on lock down. I then began trying to explain everything to my husband and what I heard. 'Do you think it's Bigfoot?' I was floored when he said that, but I still had forgotten about the fact that I prayed to see one![239] Flash forward to a month later and we had finally settled in. There wasn't any activity like that first night we moved in. We eventually met our neighbors and talked to them about the wildlife in the area. They told us that bears have been seen around our neighborhood,so Robert and I began to think that what I heard might have just been bears. That put me at ease and I began to enjoy sitting outside on the balcony upstairs and drinking my coffee."

"The peace, the quiet and relaxation was short-lived. I began to hear more strange noises like the sound of trees falling, usually when there weren't any strong winds or storms. Other times I would hear screams that were too loud for any human vocal sound to make. I started to hear a lot of wood knocks. Large sounding footsteps as if a giant was walking through the woods. I've heard what sounded like a goose or a duck being choked to death until the sound quickly went quiet. The amount of turkeys we had on the property slowly started to disappear. I also heard the sound of something growling at me while on the balcony looking out into the forest."

"So with about five months of living there on the lake, I was sitting outside on the balcony one evening drinking my coffee and then without warning beside the house I heard a cat screaming as it was being choked or squeezed. The screaming quickly stopped and I heard a small thud. The sound continued across the yard and up a tree. I then heard a loud thud, thud, thud. I thought whatever ran across the yard was a deer, but quickly realized, wait a minute, deer don't chase cats up trees! I stomped my foot on the floor of the balcony. When I stomped my foot, I heard whatever this was loudly turn around and start to walk slowly making the thud sounds again! I slowly raised up from my chair to see if I could see anything, and with amazement I noticed a large cone shaped head and shoulders, with long hair covering the neck. It was so tall. The balcony is at least 12 feet up from the ground and I could see it clearly. Immediately I began to feel sick."

"Immediately I had the worst headache I ever experienced in my life. I instantly became confused, and I couldn't speak or get my words to come out right.[234] I knew at that moment what I had just witnessed but couldn't say anything that made sense. I had gone inside and sat on the foot of our bed, my husband was just getting up and knew something was terribly wrong with me, he kept asking me over and over, 'What's wrong honey'?"

Like so many cases over the years when people are in close proximity these creatures have an ability to destabilize you in some way. Infrasound is typically considered the culprit regarding how they can do this. But I have heard this from so many people that they get an intense headache and oftentimes the only way it stops is if you ask them to stop, apologize to them verbally and the headache or the sickness feeling will subside. Otherwise the symptoms ensue for hours.

When Sherry tried to respond to her husband Robert about what was wrong, her words coming out were just not making sense, she was babbling. "The headache, oh it was horrible, it went on for hours before it finally stopped. When it did finally stop I told my husband what happened, I felt silly and hoping he would believe me, he said he did. So that day I decided to contact someone who was an expert on bigfoot. I finally found someone who was in Bowling Green Kentucky, James Wiggins."

"After I told James about my experience he asked me how I was feeling. I told him about the severe headache and the confusion that occurred and not being able to speak. He explained to me that I had been zapped. Well, I didn't know anything about being zapped, but he explained it all to me. So as time went by I recovered but still had my guard up, afraid to go outside at night but I didn't give up my balcony and coffee."[235]

The chaos would continue. They would have rocks thrown at the house during the night when they tried to watch television. The noises they would hear coming from the woods became more and more until one of the events was mind boggling. Sherry even heard two of them talking to one another, also known as “ samurai chatter” made famous by Ron Morehead and Al Berry with the “Sierra Sounds”.[236] “Once while sitting on the balcony I heard two of them chattering back and forth, one sounded like a male and the other like a female, he sounded as if he was scolding her and sometimes she sounded as if she was whimpering.”

What I have found is that empaths, clairvoyants or sensitive people can not only see spirits, but things like Bigfoot, UAP and UFOs and even the occupants of these craft. All of the strange phenomena described also notice these individuals. They behave and act in a similar way, appearing and disappearing from our reality with ease. But they are showing themselves to certain people for a reason and I don’t believe it is to keep it to ourselves because others are having these same encounters as well and they have been for hundreds if not thousands of years. The ones that overcome their fear of it continue to level up it appears. They see more. They are the messengers.

As they were closing in on their one year anniversary at being at their new home they were coming back from a typical drive to town one day in late August. They were on their way back home. She stated that it was a clear, beautiful summer day. As they rounded the corner in the road only a few yards from their driveway Sherry had glanced down to look at her phone. She quickly caught a glimpse of something red out of the corner of her eye and without looking up assumed that the leaves had already started to change color. A second later she looked up as she realized that the foliage wouldn’t be turning colors for another month. When she saw what the red color was, she lost her breath.

“And there he stood, in a split second I was angry! I was wondering who is that idiot in that gorilla suit? Then reality set in and I thought that's not a suit, it's funny how things can play out in slow motion, I had seen everything about him in probably less than five seconds but it seemed so long as if time stood still when I was looking at him seeing every little detail.

All I can say is, it's amazing, there he stood, beautiful auburn red hair, *he looked so well groomed*, his hair had an amazing sheen or shine as the sun was shining on him. His eyes were perfectly rounded and they were a clear blue/green. His brow stood out with his eyes set deep into his face. His nose looked pugged and his nostrils were flared. Its skin on its face and hands had no hair."

Sherry continued describing the creature. "The color of its face was an unusual gray. It was kind of a soft gray. The skin looked soft like smooth leather. The shoulders were very, very broad and they reminded me of a football player only larger. The legs were big as tree trunks but looked too short for his body. The arms were enormous! His hands were so big and I just remember focusing on the fact that his hair was so beautiful. He had long locks of hair dangling underneath his arms and it reminded me of a brand new mop that had never been used."[237]

What a lot of researchers dream for and which Sherry prayed a year ago which continued to escalate into what happened next. "His eyes locked onto my eyes. I realized as we stared at each other...he looked to be amazed at unexpectedly seeing me as I was of him. His arms were out to his side as he was looking at me. As we stared at each other I remembered that his arms were swaying up and down slowly as if they were in slow motion."

Witnesses discuss things happening in slow motion where every minute detail is retained.[238] Some elements could be interpreted as one thing but could be something else entirely. Witnesses have described seeing "sparkles", or a glow, some sort of haze or what could be deduced to be some sort of "heat devil" or temperature aversion when they are watching a UFO or observing a bigfoot or cryptid-like creature. Orbs or small balls of light are often seen not only at night but during the daytime as well around some of these unidentified aerial phenomena or terrestrial creatures.

Sherry continues, "I can remember there were tiny bugs flying up from the ground and moving around in the sunlight. He stood there with his legs spread apart and his arms out from his side looking directly into my eyes. Every detail seemed to be in slow motion and lasted for what seemed longer than normal. It would be safe to say he was about 7 feet tall, maybe more. I was so amazed, but I think based on the look on his face, in a weird way, he was amazed to see me too."

"As this whole scenario played out before my eyes I never thought once to lift my phone up and take a picture. I will say this, that picture of him I see in my mind will forever remain burned into my memory. I think he was a beautiful creature, but I wouldn't by any means want to approach him because he was so enormous."

"After that day I couldn't bring myself to tell my husband about my experience until roughly two weeks later. I wanted to tell him when it happened but after I was worried he would think that I was losing my mind or something. I finally got up the nerve to tell him and I was relieved that he believed me. It was months later in December that we couldn't take all the activity at this home so we decided we needed to leave. I will just say this Ronny, whatever you ask for in prayer...you may actually get it, I did."

She's right. A memory came back to me when I was on a cruise with my family about five years ago. I was alone on the deck looking out into the navy blue ocean and marveling at the fact that we were on our way to arriving in Bermuda within the hour. When I was growing up I spent a lot of time reading books about animals and cryptozoology. There was one kids book about strange animals like the platypus and others. It had two colorful pages of hand painted Flying Fish. I remember it was highlighted that they can be found in Bermuda.

As I stood on the deck overlooking the ocean I said out loud, "God, show me something really cool." It was in a matter of minutes I noticed off in the distance movement. When I strained my eyes to see, I watched five or six Flying Fish emerge from under the ocean and they were flying in the air alongside the ship twenty or thirty yards away. *I lost my breath.* I had always wanted to see them in real life and low and behold once I asked for it they appeared. It was an amazing moment and another reminder that someone seemed to be always listening and then revealing something.

What binds these stories together is not simply what was seen, but the conditions under which it appeared. Curiosity became an invitation. Prayer became permission. Sensitivity became visibility. Across Indiana, Kentucky, and beyond, the phenomenon did not arrive randomly, it responded. These encounters suggest that whatever moves through forests, skies, and thresholds is not merely observed, but engaged, and that once engagement begins, it rarely confines itself to a single witness, a single moment, or even a single generation.[240]

Chapter Seventeen

ASCENDING DURING THE AGE OF DISCLOSURE

By this point in the book, the question is no longer whether these experiences are real, but what conditions allow them to unfold—and why they cluster where they do. Chapter Seventeen moves beyond testimony and into pattern recognition, examining how belief, geography, consciousness, and intention may act as catalysts rather than coincidences. What emerges is not a mythology of monsters, but a framework for understanding how interaction occurs when thresholds are crossed.

If there is a single common denominator threading ancient encounters and modern disclosures, it is light behaving intelligently, Not light as illumination, but as an interface that is capable of movement, response and apparent awareness. This allows us to look at contemporary physics not as a replacement for myth, but as a new language for something humanity has always encountered and will continue to encounter in the future.

Dr. Salvatore Pais, a well-known aerospace engineer associated with the U.S. Department of Defense, has either intriguing ideas about plasma and its potential as a form of conscious life or he knows more than what he is saying. His interviews, particularly the one in "Salvatore Pais a Brilliant DOD Engineer discussing the Plasma & Consciousness Connection," delve deep into the relationship between advanced physics concepts and our understanding of consciousness.[241]

"What if there are plasma beings, a whole other form of life, living amongst us that we haven't even realized until just recently?"

Dr. Salvatore Pais

Pais's ideas are revolutionary, suggesting a radical rethinking of what we know about life forms, specifically hinting at the possibility that plasma-based entities could exist among us without us realizing it. Pais begins by framing plasma, typically thought of as a state of matter like solids, liquids, and gases, as something more significant—something that might hold the key to understanding different forms of consciousness.

Plasma is essentially a hot, ionized gas composed of ions and free electrons, and it makes up the majority of the visible universe, including stars and interstellar space.[243] Pais suggests that if this form of matter could somehow be structured in a specific way, it could perhaps attain a state of consciousness, raising the question of whether intelligent plasma beings might exist.[244]

One of Pais's core arguments is that life and consciousness may not be limited to the biochemistry of DNA and cells. Instead, he proposes that other forms of life, such as plasma beings, could have evolved based on fundamentally different principles. If true, this implies that entire ecosystems of these beings might have been coexisting with us for millennia without our awareness, simply because they exist in forms we do not have the sensory capabilities or technological tools to detect. Pais further explores the idea of the "conscious universe."

He aligns his discussion with speculative ideas in quantum physics, where consciousness is proposed to interact with the fundamental building blocks of reality. He suggests that plasma, with its energetic and interactive nature, could act as a medium that embodies some form of awareness. This line of reasoning takes influence from advanced physics theories, like those involving quantum fields and energetic connections throughout the cosmos.

If plasma entities do indeed exist, they could potentially interact with the physical world in mysterious ways that might explain phenomena such as UFO sightings, ghostly apparitions, orbs, or strange balls of light including other anomalous events often considered supernatural.

In the context of advanced propulsion and technologies, Pais's theories could also serve as a foundation for explaining many of the "unconventional" patents he has filed in the past—patents that include room-temperature superconductors, high-energy electromagnetic field generators, and even aircraft capable of extraordinary feats.[242]

These inventions, although theoretical, provide a glimpse into a future where our understanding of the physical universe is dramatically expanded, including the existence of plasma beings that manipulate or utilize such technologies. Pais also touches on the philosophical ramifications of recognizing plasma beings. If life exists in other forms beyond what we can see, touch, or hear, what does that mean for our place in the universe?

Pais argues that this would necessitate a reevaluation of humanity's perspective on life, consciousness, and the cosmos. It could lead us to a broader, more inclusive understanding of what constitutes sentient life and how we interact with our environment. Dr. Salvatore Pais introduces a perspective that bridges the gap between hard physics and speculative metaphysics.

In September of 2024, recent observations detected equatorial plasma bubbles (EPBs) over the pyramids of Giza.[247] Of all places around the world, this to me screams significant.

These plasma bubbles, which are pockets of superheated gas, can impact satellite communication and are usually observed at low latitudes.[248]

The significant part of this discovery is that Chinese scientists detected these bubbles from over 8,000 kilometers away using a radar system called LARID, located on Hainan Island. This new method provides valuable insights into space weather monitoring, which could potentially help mitigate disruptions caused by these phenomena in the future.

The number of these plasma events increases with solar activity and can vary from season to season just like the weather. Like dark matter, which makes up 95% of the known universe, we still don't have a clue as to what it really is? We can also speculate that these plasma bubbles may be integral to understanding the fabric of the universe on a deeper level and maybe a link to how interdimensional travel may be possible for some of these beings.

Dylan Borland, U.S. Air Force veteran, who testified before during the September 9, 2025 congressional hearing on *Restoring Public Trust Through UAP Transparency and Whistleblower Protection,* worked with special operations teams and while living in the barracks on base in 2012, witnessed something otherworldly.[249] Here's what he had to say.

"There was a little smoke pit outside, I was there on my phone looking across at the flight line and I saw a white light pop up and stop about 100 feet in the air. I thought it was a weather balloon as I have seen tests from there before, weeknight, normal thing, not surprising. I finished my cigarette and I began walking up towards the flight line. There was a track because I had done 3 months of night work and it had been weathered down.

As I began walking on this track towards the light, the light then flew across the base and across the flight line and as it flew to me a triangle manifested around the light. I can't tell you if it is active camouflage and I can't tell you if it appeared around the light...but I can tell you that it was a white light and then it was a triangle.

It stopped about 100 feet in front of me and approximately 100 feet above me. My telephone got extremely hot and completely froze dead. I remember how thick it was. It was about 1 to 2 stories thick and an equilateral triangle and I could never see the top of it and the edges were 90 degrees. There were four lights in total. One light on each corner and a larger light in the center about two to three times the size of the corner lights, but what was really odd was the outside, the best way to describe it is looking at a James Webb telescope picture where you have all the colors and the black background, so the craft itself was this black metallic flight paint but on top of the craft was this gold, lava, plasma type fluid going over and around the craft.

I am under this for about two to three minutes and then the center light flashes two to three times, no sound and immediately shoots up to commercial jet level minimum, in my opinion, and I immediately feel static electricity all of my body and then i smell the smell of after a thunderstorm or lightning storm, that really strong thunderstorm level.

It gets up to flight level and I am trying to get my phone reset and I can only see the center light at this point. If I didn't see it take off, I would think that it was a star. It hovers up there and slowly begins to move due east out over the Atlantic Ocean. I finally got my phone reset. From the entirety of seeing the light and until it took off up into the sky was about 15 minutes."

When he started to talk with others on the base, some of the older officers told him that he should just leave it alone and stop talking about it. He then endured years of problems from different agencies and essentially lost jobs and is currently unemployed because of it. He was a courageous whistleblower to come forward and one that also has indicated that these craft are doing things that we don't currently have the capability to perform its maneuvers, as well as its shape-shifting and cloaking capabilities.[250]

The craft also looks like it responds to Borland recognizing him on a conscious level. It was aware of him watching it and moved over to his position. His description of it going from a ball of white light and turning into a triangle craft and using his term "manifested" sounds like it shockingly appeared, maybe from another dimension. The gold plasma swirling on and around the craft sounds like the material that makes the orange orbs as well as the Thunderbirds my wife and I witnessed.

There is a brilliant man by the name of Dr. Salvatore Cezar Pais, a Romanian-American aerospace engineer and inventor associated and working directly with the US Space Force, whose work likely is out of this world as well. He previously worked with the US Navy. In fact, Pais holds at least 6 to 7 known U.S. patents, mostly filed between 2015 and 2019, all of which are assigned to the U.S. Navy's Air Warfare Center Aircraft Division. Receiving his PHD from Case Western reserve for NASA sponsored research.

These patents include futuristic ideas:

- **Inertial mass reduction device**
- **Plasma compression fusion device**
- **High-frequency gravitational wave generator**
- **Electromagnetic field generator**
- **Room-temperature superconductor**

These patents have been affectionately called the "UFO patents". Plasma is the fourth state of matter after solid, liquid and gas. What we know at the moment which is subject to change is that plasma is a superheated, ionized gas where electrons are stripped away from atoms, creating a soup of free electrons and ions. It behaves differently from other gases due to its highly electrical conductive, magnetic field-responsive, and often glowing properties. The sun and stars are giant balls of plasma. Auroras and lightning bolts are made of plasma and we also have commercial products like Plasma TVs and Neon signs. But can plasma be conscious?

In 1928, Irving Langmuir who was an American physicist and chemist coined the term “plasma”.[245] He named it as such because it reminded him of blood. The ionized gas in these tubes behaved like a fluid that carried electrons and ions, similar to how blood plasma carries cells in the body. Langmuir believed that this plasma could be capable of life. He believed that the plasma was not only capable of action but intention, which is a theory of consciousness. Dr. Pais postulates that plasma could very well be a type of life form.[246] Some of these UAPs and craft could very well be alive like we live and breathe.

Carl Jung said “the decisive question for man is: is he related to something infinite or not?”[251] On the day of Pentecost, the apostles were not struck by lightning, nor consumed by flame. The room did not ignite. No structure was damaged. Instead, something luminous appeared, divided itself, and positioned over each individual present. The Book of Acts says the fire did not fall. It hovered. That single detail—often skipped over—changes everything.

The original Greek is precise. The phenomenon is described as glōssai hōsei pyros diamerizomenai—tongues, as of fire, distributed. And then the text adds a strange, almost technical phrase: It sat upon each of them.[252]

Immediately after, cognition changed. Fear vanished.[253] Speech transformed. Identity expanded. A group of people who had been hiding behind locked doors stepped into public view with absolute certainty. If this account were written today—without religious framing—it would not be called a miracle. It would be logged as a contact event. Ancient people did not have language for plasma, electromagnetic fields, or non-local intelligence. They described what they knew.

Growing up in a Christian home there were elements of Sunday School or church that didn't make sense. Concepts like the Holy Spirit, "speaking in tongues" when a person begins speaking other languages in an instant, hearing the words coming out of their own mouths for the first time as if they were downloaded. Or with their hands up in the air pointing to the heavens during praise and worship. It just looked odd and seemed weird coming from Catholicism. This is one of the things that differentiates being a Pentecostal from other religions like Catholicism or Methodist. I now consider myself more Buddhist / Christian than anything else and more of a spiritual person than a religious one. In hindsight I do believe that all of the paranormal things in my life happened shortly thereafter which some would consider "mystical experiences".

I have had the strange experience of being "slain in the spirit" two times. Once while at the Assemblies of God in Leominster on a Sunday morning and another at a bible retreat in Rhode Island in the evening, both when I was a teenager. A preacher or someone that is considered "anointed" prays over you and typically there are several people on either side and behind you for what happens next. I went up to challenge it as I didn't believe what I saw when people were being prayed over. They would essentially pass out or black out falling into the arms of the deacons or church members behind them. Slowly waking up but feeling different, better now than before the experience. It's like getting unplugged and being plugged back in.

I walked up defiantly waiting in line and watched as one by one people were going down left and right after being prayed over. I was thinking that I was going to prove them wrong, that my mind was stronger and that these people were weak or simply just faking it. It was my turn and the pastor laid his hands on my head and began to pray. He started to shake or push my head in an aggressive manner with each word like he was trying to wake me up. I thought to myself, "nice theatrics buddy."

Then it started to happen, I could feel this overwhelming feeling of energy coming from his hands. I remember trying to fight the feeling as the pastor was praying over me and then "boom" I just blacked out and lost all control of my body falling backwards into some strangers arms who were ready behind me. Like the others before me, they slowly lowered people down to the ground and gently moved them aside and prepared for the next person.

I woke up a few minutes later lying on the ground in a weird way where I could hear the distant talking of people around me while slowly the volume of my surroundings turned up like I was emerging out of water, the sounds swelling back into my ears like someone was turning up the volume button from 0 to 10.

I couldn't see anything and out of nowhere from the inky blackness of zero vision, the lights, shadows and figures around me slowly turned into lighted reality. My eyes, almost like apertures on a camera, were being turned on and everything slowly merged into visual reality with audio. I was back. But where did "I" go? I didn't have any memory of what was happening in between the experience of walking up and then waking up.

The feeling was unlike anything I have ever experienced before. Refreshed and recharged it was like a spiritual tune up. I had an immediate sense of power, intense feelings of love and completeness and a total lack or void of fear. Simply put, clarity. It was real. The energy I was aware of, wherever it was coming from was palpable. God was real or whatever this was...was real. Some would equate this awareness as "becoming = Born Again".

The other experience in Rhode Island happened several years later in a church as well during a lightning storm no less and the power went out when the pastor said that Devil was trying to stop the evening from going any further. This time I knew what I was in for and was ready this time. With the emergency lights on in the church, the pastor didn't stop as people were being slain in the spirit in the same fashion. I had a different approach this time, I wasn't afraid I was curious if it would yield the same result.

After I awoke from this experience I remember being unafraid and walking back to our cabins where we were staying in pitch darkness and somehow made it back with no aid of light and others following behind me amazed at how I was somehow 'seeing in the dark".

While writing this book the caricature image of flames of fire over peoples heads from a children's book we had at the church that I recalled reading entered into my mind's eye. This happened a few days in a row so I started to look into it more as I thought that this sounds like the orange orbs or balls of fire we are seeing in modern day.

The major occurrence of this comes from the Book of Acts and talks about the Tongues of Flame at Pentecost where the "Holy Spirit" descended down on Jesus' apostles like tongues of flames. Today would be described as glowing, plasma-like lights. "A sound like rushing wind filled the house and divided tongues as if made of 'fire' appeared and rested on each of them". The tongues of fire would separate and hover over the apostles' heads. The phenomenon was visual, audible, localized and luminous as if burning fire but not. Like the orbs of modern day. The "Chariots of Fire" would be another example or description of what they were seeing, trying to explain something foreign or alien in today's terms.

You also have the story of "Moses and the Burning Bush" which never made sense to me. A bush engulfed in flames and not being consumed and communicating as Moses hears a voice from within it. Was this audible in his mind like telepathy? An important detail is that the Hebrew word often translated as "fire" (esh) can mean radiance, brilliance or an energetic presence that wasn't an ordinary fire. This matches an apparent intelligence and self-sustaining luminous object or phenomena. Across the Bible you will also see Ezekial describing "whirling luminous wheels" within wheels. Elijah ascends in a fiery vehicle and the Israelites follow a pillar of fire that moves, stops, and guides them. It looked like a cloud by day and fire by night.

Are these all potentially plasma based luminous phenomena, interdimensional lights that are conscious and which God or the creator is communicating to humanity through? Sounds like what modern day experiencers are seeing today with anomalous light encounters in the forest, in their homes and in the sky. The phenomenon presumably is as old as time itself.

Now let's look at 3iAtlas that is approaching with December 18th being its closest to earth. It is a mere seven days from Christmas and I instantly think of the Star of Bethlehem as being something other than a star as well. It just happens to be leading the way of the three wise men or Magi who followed this "star" which then hovered over the birthplace of Jesus Christ. I also discussed the idea in my last book that Jesus is considered by many, who don't believe he was necessarily the "son of God", but he was an Ascended Master.

The early Christian writers also didn't believe that the 'star of Bethlehem" was a normal star. Ignatius of Antioch said "the star outshone all others and caused confusion among the stars."[257] The star moved like an "invisible power, John Chrysostom.[258] If we were to describe the Star of Bethlehem today in modern neutral language devoid of theology or labels, a luminous intelligence object that suddenly appeared, maneuvered with intent, guided special observers and stopped at a specific location then we would unequivocally call it a UAP or UFO.

It gives the impression to have been a plasma-like orb or consciousness similar to the other "fiery-like" objects depicted in the Bible. We hear from Rep. Anna Luna on how Enoch and the books of Enoch have something to do with the UFO enigma. Atlantis surfaces at the top of these discussions and conversations as well with the emergence, confirmation and disappearance of these various objects in and out of our oceans.

The Book of Enoch describes "the Watchers" not being from another planet but crossing a boundary and descending down from a higher realm.[256] They came teaching forbidden knowledge, altering human development which is believed to have produced unintended consequences. Heavens stacked upon heavens, portals, gates and luminous beings with the reality structured in dimensions of consciousness.

What Enoch describes aligns with what we are being told now that we are seeing objects that are showcasing nonlocal behavior, they don't obey inertia or modern physics. They break the rules. The objects disappear and reappear and interact with one's perception and awareness. The beings Enoch encountered appear as light and shift form and move between "layers" of reality" with light as the medium of transition. These light beings communicate via mind or dream and induce a sense or feeling of awe and fear but not violence.

We have moved the label from UFOs to UAP, from extraterrestrial to non-human intelligence and psionics, telepathy and consciousness being a part of the process as well as the makeup of these "craft". It is the same structure across time just wearing different clothes. When these lights seem to intensify around the world in the past they were peak moments for humanity. Are we approaching a tipping point moment for human consciousness? Are we about to approach another boundary or threshold that we will break through? Think of where we are now with AI or Artificial Intelligence, what some people are referring to Alien Intelligence. Will it unify humanity or break us apart even further?

Reflecting on my experience of being "slain in the spirit" as well as the stories from the book of Acts with the tongues of fire, I recall several times that during the middle of the night I would awake to see a small ball of fire appear and disappear. It would always be to my right but I could never see it straight on, it would come into view but then vanish when I tried to look at it. This ball of fire would light up the room. Strangely, I believe that these occurred *before* being slain in the spirit. The feelings of being recharged or rewired would subside within a few days and then within a week or so I would feel back to "normal". The veil or covering would be lifted but it slowly morphed back on. So it is a state of mind that needs to be maintained or reengaged.

"Our normal waking consciousness...is but one special type of consciousness, whilst all about it, parted from it by the filmiest of screens, there lie potential forms of consciousness entirely different."[259]

- William James

What's intriguing is that children and shamans occupy similar cognitive states where there is a reduced top-down filtering and a reduction of self or ego, preconceived notions and biases that cloud one's mind. Neuroscience and Anthropology highlight that the mind has the capability to access the same entities via trance-like states, illness or a cultural ritual, that are seen throughout time and cultures as either spirits, angels or alien beings from another place.[260]

So in essence they are not seeing more of this world they seem to be suppressing and seeing less and now have the ability to perceive across layers most adults unconsciously remove or delete. They are turning into another frequency. Contact not only changes the environment but more importantly it transforms the perceiver into a receiver.

Looking back at the transformation of the apostles after the tongues of fire hovered over their heads there were several profound things that happened instantly internally and then rippled outward. The fire over their heads signified an activation as the apostles were filled with the Holy Spirit embodied with a divine presence that instilled courage, clarity and a lack of fear. They also like in the Matrix when Keanu says "I know Kung Fu", they were downloaded with the ability to speak other languages and understood by others. And on top of that I believe that telepathy was activated as well because everyone that was in the room around them was able to understand languages that were of the listeners or perceiver origin!"

"Each one heard them speaking in his own language." Acts 2:6.

This reversed what had occurred at the Tower of Babel when humanity spoke one language, maybe telepathy and the consciousness connection amongst all humans translated them all at this time which was then collapsed into a communication confusion and the languages then fragmented reality.

Some accepted this awakening and some did not. The crowd that was there at the Pentecost event were stunned and some divided that what they were witnessing was that the apostles were drunk. The pattern of Moses with the burning bush, Elijah with fire from heaven, the experiences of Enoch and the Apostles had the fire from within transformation. Across indigenous and shamanic traditions, fire hovering over or entering the head is a sign of initiation.[254]

For example, Siberian shamans describe flames descending into the crown. Amazonian ayahuasca initiations include visions of serpents made of fire and the Celtic druids spoke of Awen, a flame of inspiration that would enter the poet or the seer. In yogic traditions and systems Kundalini is the dormant energy at the base of the spine that rises upward, ascending through chakras igniting the throat and effecting speech and vibration eventually activating the crown which is considered the divine connection or receiver. In ancient Jewish thought, fire is consciousness in its pure state.[255] This all sounds very similar to the Apostles transformation. The common thread is consciousness and the first entering through the crown or on top of the head.

"Tongues' in Greek (glossal) means languages, flames, organs of vibration. The apostles weren't just speaking words they were now transmitting a coherent and resonant frequency that others in the room understood. The light contained the coded information, the sound was the carrier wave and then consciousness translated this into vibration on an individual level. Is there a new awakening upon us? Every era before had had fire or light phenomena that connected and communicated and we are seeing it today with whistleblowers from within these hidden UAP programs, advances in psychedelic research and the state of AI rapidly accelerating and the peeling away of the layers that are covering up the truth of our reality.

The modern-day whistleblowers are discussing telepathic communication, floating luminous orbs, fire-like plasma, objects splitting into smaller lights that respond to consciousness or intent and have the ability to alter one's perception and cognition. Like Elijah, Enoch is taken alive and led by beings whose bodies resemble burning fire, faces like the sun, limbs polished bronze. He sees thrones of crystal and flame and rivers of liquid fire. These entities are energy and eventually Enoch transforms into a body made of light, he is renamed Metatron and no longer is visiting the luminous realm but now belongs to it. Teilhard de Chardin famously said, "We are not human beings having a spiritual experience. We are spiritual beings having a human experience."[261]

What disclosure now appears to be revealing is not merely the existence of non-human intelligence, but humanity's evolving capacity to perceive it. Across scripture, shamanic initiation, whistleblower testimony, and physics, the same pattern repeats: light as intelligence, consciousness as the key, and transformation as the outcome. The final threshold is not technological, it is *perceptual.*

Chapter Eighteen

THE PERCEPTION SHIFT

With the upcoming book *Bigfooters and Scientific Inquiry* based on the Cardiff University study by Jamie Lewis and Andrew Bartlett about to be published, the timing couldn't be better to highlight that we are witnessing a paradigm shift happening before our eyes not only when it comes to UFOs and UAPs but also Sasquatch or Bigfoot.[262]

The release of the most downloaded documentary in Amazon history *"The Age of Disclosure*" by Dan Farah is making waves along with the documentary *"American Sasquatch"* by David Paulides shedding more light on the strange side of Sasquatch and its connection to the UFO enigma. I have faced a lot of opposition since 2010 with me trying to connect Bigfoot, Orbs and UFOs but more are coming over to this side of the fence.[263] [264]

James Fox's forthcoming revamped and updated documentary *Moment of Contact: New Revelations of Alien Encounters* along with his body of documentary work is also cementing the idea that not only are UFOs real, that they have been interacting with humanity for eons but that we have had crashes *and a lot of them, their* pilots included. Credible witnesses have also had telepathic communications along with physical interactions with these beings. James simply texted me the other day stating, "Get ready, there's more to come..."[265]

Around the corner Steven Spielberg's new movie *Disclosure Day* about UFO and UAP Disclosure is about to shatter minds and what some may consider to be his greatest cinematic as well as cerebral work yet as he focuses on life's greatest mystery, are we alone?[266] A recent billboard in Times Square emerged as the first promo of the film with an alien at its center stating "All Will Be Revealed" in June 2027. This is all very exciting for me because I sit at the crossroads of both of these particular passions as well as consciousness studies that have shapeshifted into obsessions all of which I firmly believe hold the key to our reality.

The Cardiff University of Wales sociological study which is the center of the upcoming book, examines Bigfoot researchers as a scientific community operating outside traditional academic structures.[267] Through interviews, field ethnography, and analysis of Bigfooting practices, the authors conclude that Bigfooters engage in forms of investigation that closely resemble scientific inquiry—casting tracks, analyzing audio spectrograms, comparing ecological data, and debating competing hypotheses. I have always been curious and fascinated by science but I have always been someone that goes against the grain, swims against the current not because I want to resist the establishment but because I follow my gut or my intuition and I have always believed I am protected and directed.

These researchers, these curious minds are not anti-science; rather, they actively seek scientific validation and often collaborate quietly with academics. The same can be said for UAP study and research and those boldly going and literally risking their lives for the truth which has been systematically compartmentalized and highly classified, its secrecy stuffed into a dark hole hidden from the general public.

There are moments in the history of science when the real breakthroughs do not come from laboratories or universities, but from the edges—from hunters, wanderers, skywatchers, thinkers, inventors and people stubborn enough to keep asking questions that institutions refuse to touch. The search for Bigfoot has always lived in that borderland, and for decades, the people working in those shadows have been dismissed as amateurs, eccentrics, or worse. But a closer look at how knowledge is actually built tells a different story.

Bigfoot research has always been a story of independent minds refusing to wait for institutional permission. It is a story of the people who spend time in the woods, who listen deeply, who collect traces and fragments and stories, trying to piece together the shape of something vast that science has not yet allowed itself to acknowledge. More often what ignites this inquiry is a personal experience that they can't explain. These are not the people standing in the way of science—they are the people expanding its borders.

However, Bigfoot inquiry lacks the stabilizing structures that define scientific disciplines: shared standards of evidence, peer review, formal training, and agreed-upon reference materials. In the absence of a confirmed body, DNA, or consensus benchmarks, Bigfooters face what sociologist Harry Collins calls "experimenter's regress," where evidence cannot be validated without authoritative standards and standards cannot emerge without validated evidence.[268]

This leaves Bigfoot inquiry highly vulnerable to hoaxes, internal disagreement, and fragmentation. Outsiders—ranging from hunters and foresters to audio engineers and former military personnel—bring valuable skills and novel perspectives that mainstream science might overlook. Yet without a disciplinary framework, these contributions rarely crystallize into cumulative knowledge.

Ultimately, the authors of the Cardiff study argue that Bigfooters inhabit the "borderlands of legitimate science": a hybrid space where scientific reasoning thrives, but institutional science is absent. The study concludes that while independent and rogue researchers are essential for bringing new insights to anomalous fields, the lack of organizational coherence prevents these insights from becoming established science.

But is this because Bigfoot doesn't neatly fit into the box of science?

The work suggests that independent, rogue, and cross-disciplinary investigators are vital engines of discovery, yet their insights require institutional structures to achieve lasting scientific impact. Bigfooting resembles a pre-paradigmatic science, where no single model dominates and competing schools of thought operate simultaneously.

This matches the study's conclusion that Bigfooting is fractured and lacks consensus.

Paul Feyerabend argued that progress sometimes requires breaking rules ("anything goes"). Bigfooters embody this. UFO researchers embody this as they innovate outside academic constraints, use hybrid methods, and question dominant assumptions.

Charles Fort said science maintains itself by ignoring inconvenient anomalies. Independent researchers operate precisely in this rejected zone—collecting anomalous data mainstream science won't touch.[270] Outside researchers are the early detectors of anomalies and they are the engines of innovation.

But the real insight of the study lies in how it explains the role of the outsider. In mainstream disciplines, research is shaped by what Thomas Kuhn called a "paradigm"—a shared model that tells scientists what counts as real, what questions are acceptable, and what methods are allowed.[271] Bigfoot, by its very nature, falls outside those boundaries. It is an anomaly, and anomalies tend to be exiled until a discipline is ready to deal with them. Kuhn argued that these anomalies often sit ignored for years until they force a paradigm shift.

That is exactly what is happening NOW.

Bigfoot researchers—foresters, hunters, wildlife trackers, audio experts with military backgrounds, and curious civilians—occupy the gap between lived experience and academic legitimacy. Academic primatologists and biologists cannot explore Bigfoot openly; the professional risk is too high. So the responsibility falls to the outsiders. An example is "Project ZooBook," where credentialed scientists meet quietly with Bigfoot researchers behind the scenes, unwilling to risk their reputations by speaking publicly.

This says something profound:

The search for Bigfoot has become a refuge for scientists who still want to pursue the unknown—just not under their own names.

And it echoes a pattern we've seen again and again in history.

Meteorites were once dismissed as superstition.

Gorillas were a "mythical" creature until 1902.

The platypus was mocked as a hoax.

Continental drift was scientific heresy for decades.

UFOs are not real.

We are alone in the Universe.

Each breakthrough began with an outsider pointing to something the academy wasn't ready to see.[272]

The study concludes that Bigfooting lacks the stabilizing mechanisms that help scientific communities build consensus—shared standards, peer-review, agreed reference materials, and professional training pipelines.[273] With no accepted body, no verified DNA, no universal benchmark, Bigfooters face what sociologist Harry Collins calls "experimenter's regress": they cannot validate evidence without a standard, and they cannot create a standard without validated evidence. It's a scientific Catch-22.

But despite these limitations, the study shows that the field moves forward anyway—driven not by institutions but by passion, stubbornness, and a collective intuition that something is out there. The heart of it all is the idea that discovery often begins with people on the margins, the outsiders, the punks of the paradigm shifts. I am one. You are one too.

And that pattern—of outsiders carrying the torch while institutions look away—is a theme that runs through every chapter of the Bigfoot and UFO mystery. It is what allows the phenomenon to evolve, adapt, and persist. It is what keeps the search alive. And it is why, when the day finally comes that Sasquatch steps into the light, it will almost certainly be one of these so-called "amateurs" who finds it first. People like Daniel Lee Barnett and the new wave of researchers.

Because history shows that the edge—not the center—is where discovery happens.

The Intelligence Behind the Mask

If Bigfoot and UFO research both sit on the borderlands of science, then consciousness is the territory beyond the border—the dark forest at the edge of the map.

For decades, mainstream science has treated consciousness not as a frontier, but as an inconvenience. A complication. A word to be avoided in grant proposals. A puzzle that refuses to fit inside equations.

And yet, it is the one variable that keeps showing up everywhere phenomena occur.

People who see UFOs report time distortions, telepathic impressions, and synchronicities that cascade for days afterward.[274] People who encounter Bigfoot describe "panic blasts," overwhelming dread, silence zones, and impressions of being observed with an intelligence that feels non-human yet aware.[275] UFO experiencers speak of orbs appearing inside bedrooms; Bigfoot witnesses see orbs in the woods moments before or after encounters.

These patterns are not separate because they are connected through human experience and that experience is consciousness, the forbidden link — the one that science has tried to ignore, but can no longer escape.

Every case study tells a story that challenges our conventional understanding of reality. In the forests of North America, researchers have documented Sasquatch sightings that share uncanny similarities with reports from the Scottish Highlands, the Amazon, and the Russian taiga. Witnesses describe the same behaviors: silent observation, ritualized knocking, and a sense of intelligence that transcends instinct. These are not random anomalies—they are patterns, repeated across cultures and centuries, hinting at an intelligence that exists just beyond the edge of perception.

Cross-cultural accounts further strengthen this narrative.[276] The Green Man of UK folklore, North American "wild men," and shamanic forest guardians all describe beings that mediate between worlds, offering both danger and insight.[277] Similarly, indigenous interpretations of UFOs often frame them not as machines, but as manifestations of intelligence beyond human scope, capable of teaching, testing, or communicating. In combining these threads—Sasquatch encounters, UAP reports, and consciousness research—we start to see a pattern of interaction between dimensions. The knocks, the sightings, the electrical anomalies—all of it points to an intelligence operating within the subtle matrix of our reality.

This hybrid framework allows us to see the forest, the sky, and the unseen spaces in between as part of a continuum. Every encounter, every report, every unexplained phenomenon becomes a data point in a larger map of consciousness, one that spans biology, physics, folklore, and personal experience. By embracing storytelling alongside empirical research, we gain not only understanding but a framework for dialogue—with the forest, with interdimensional visitors, and ultimately with the limits of our own perception.

We are seeing the same patterns and behavior. These parallels are too symmetric to ignore. The behavior doesn't match typical animals. It doesn't match typical craft. It matches intelligence interacting through different modalities.

The convergence is not a theory but the outcome of patterns repeating across thousands of cases.[278]

Bigfoot is known for:

- appearing and vanishing rapidly
- inducing fear or euphoria
- leaving behind trace evidence
- interacting selectively with witnesses
- mimicking human speech
- creating structured formations
- emitting infrasound
- observing without being seen

UFOs are known for:

- blinking into and out of view
- inducing fear or euphoria
- leaving behind trace evidence
- interacting selectively with witnesses
- mimicking aircraft behavior
- creating structured formations
- emitting EM effects
- observing without being seen

At the center of the convergence is something startling: The reality of it is, the phenomenon adapts to the observer. Like in the famous double-slit experiment where light can be a particle and a wave at the same time and only changes if the observer is present, or that the CIA has declassified the documents stating that were are living in a holographic simulation then our reality as we know has been classified.[279] [280] The emergence of AI is a tipping point because reality is heading into a place where it is indistinguishable from video games or artificially generated content.[281]

When you're in the forest, this intelligence appears as a wildman or strange lights. If you're in the sky, it appears as a craft or a fiery thunderbird made of plasma. If you're in a transformed state, it appears as light, or geometry, or consciousness. This is not deception, but translation.[282] This intelligence or "they" are coming into our reality from some other place whether that be the mysterious dark matter and energy that we know nothing about or from another realm about which I believe is a higher dimension or dimensions.

Anomalous Pioneer Jacques Vallée called it the "control system" — an intelligence presenting itself in culturally and psychologically appropriate forms.[283] That's why we are seeing similar bigfoot-like creatures in other countries that resemble one another.Think about even the deity and gods that others throughout the world worship or pray to, they are variations of the same mask. Other researchers describe it as an interactive field, a liminal presence, or a consciousness-based intelligence.[284]

Bigfoot may not be "shape-shifting," but the phenomenon behind both Bigfoot and UFOs might very well be. If a stem cell can become any type of cell, if the elements in the periodic table can shift into other forms of elements and go up the chain like the alchemist of ancient times knew how to do, from lead to gold, then this is the new paradigm: not creature vs. craft, but intelligence vs. perception. Does this also mean that we as a human race will ascend or transform into something else?

This is where the frontier is heading. And in 2025, we are witnessing the convergence become mainstream through documentaries, disclosure, experiencer testimonials, and the collapse of old stigmas.

This is not the fringe anymore.

This is the future.

"All that we are is the result of what we have thought. The mind is everything. What we think we become"[285]

- Buddha

Chapter one of *Interdimensional* starts with Amy and I in the garage. I couldn't wait to move into the new house, only a few weeks away. Her leg was touched 3 times and we eventually walked out of the garage and both witnessed the Thunderbirds along with a green UAP.

I was supposed to finish *Interdimensional* back in June. I struggled to find the time to get it done, the best way I could describe it was that I was getting some cosmic pushback. Amy told me, the timing isn't right, it will come - maybe something needs to happen first.

Yesterday was December 23rd, 2025, I was working on my citations for the last chapter. The book is almost completed. I wrote everyday for ten days and this was a fifteen hour day. For some reason I was unable to complete it, writing started at 5: 30 am and now it was almost 10: 00 pm, the citation numbers were being misplaced and I just chalked it off to me being tired. Just about to complete the book...I could see the finish line.

With Christmas a few days away, Amy had remarked that we had a big box of tissue paper somewhere in the house and she couldn't find it. "I must have left it at work." She came home when I was trying to finish up the book and was frustrated as it was still missing. I was like, "Wow, I haven't seen you this obsessed with something so trivial as tissue paper, we can buy more". She wouldn't let it go.

Amy walked into our now garage, trying to locate the tissue paper when she kicked a box and it tipped over, spilling out some of the contents. It wasn't the tissue paper but an old folder and a tin box that she had recovered in Worcester, MA. The building she worked in had a crew demoing a space for a new build. Amy, being the lover of historical things, saw photos and other items in the contractors dumpster one day at work and jumped in to recover the discarded items. Among the items were old photos of Worcester buildings, costume jewelry, a tin box and old three ring binder. Amy put the recovered items in a box and took them home where they ended up, guess where... in the garage. She had totally forgotten about it as this was over a year and a half ago. She then brings the newly discovered items upstairs to my office. I was glued to the screen typing away when she began to pull out its contents.

As we opened the folder we observed what appeared to be an autograph book that a Dr. Moses Eisenberg had compiled. The pages contained some of his personal letters from 1940 but also some as old as 1880 that he had collected. I pointed to the folder and said that "there is something alien or UFO related in that binder".

If we consider trees being conscious, Amy-looking for tissue paper, finds the letters. One of them was dated January 3rd, my birthday. That got my attention. We began to research the various names that were in this binder. We startlingly realized that these letters and/or correspondence - they were all authors, philosophers, poets or scientists...then synchronicity didn't tug on my sleeve, it punched me right in the face.

The first one was Felix L. Oswald (1845-1906) who was a German-American philosopher, naturalist, and early freethinker who challenged orthodox science and religion, arguing that human consciousness, nature and intelligence extend beyond materialist expectations. The letter we were now holding delicately in our hands was dated 1896. Oswald openly speculated about unseen intelligences, evolutionary mind, and non-human agency operating through natural law!

The Plurality of Worlds was a common 19th-century view that was shared by many Victorian scientists and philosophers that the universe was teeming with life, humanity is not unique or central and that other planets likely host intelligent beings. This was considered rational cosmology, not fringe belief in the 1800s.

He wrote about nature as a living intelligence, evolution as a conscious process and mental or moral advancement beyond humanity. He leaned toward naturalistic pantheism, spiritualized evolution and human consciousness expanding towards higher forms like a cosmic mind.[286]

Oswald believed that nature evolves intelligence organically where Tesla believed that the universe behaves like a vast energy field, possibly responsive to mind. Oswald provides the philosophical permission for Tesla's idea that intelligence may be embedded in the fabric or reality itself.

Oswald was a quiet historical "witness", someone who was writing a century earlier who independently arrived at the same intuition that the book Interdimensional traces: that humanity has long sensed an intelligence beside us, not above us, and not fully physical.

His work *Days and Nights in the Tropics,* written in the early 19th century, was his firsthand naturalist account of Central America where prolonged immersion in unfamiliar ecologies led him to question strict materialism and suggest that intelligence may be embedded in nature itself, not just human minds.[287]

His work expresses that sustained exposure to liminal environments like jungles, heat, isolation, altered rhythms has historically catalyzed insights about non-human or non-local intelligence. He was listening to the tropics in the 19th century and I am listening to modern anomalies now, we both arrived at the same conclusion from different centuries.

Leo Lesquereux (1806-1889) was a Swiss-American botanist and paleobotanist who was closely associated with Harvard through his collaboration with Louis Agassiz and is considered one of the founders of American paleobotany.[288]

He studied fossil plants and ancient ecosystems, helping establish that Earth's deep past preserved patterns of intelligence and order long before humans appeared. He believed that intelligence and memory are embedded in nature itself. It is written into stone, coal seams, and ancient forests waiting to be ready by those willing to look across time.

Another letter was from Robert H. Lowie (1883-1957) who was a foundational American anthropologist who documented indigenous cosmologies with unusual restraint, emphasizing that so-called "primitive" beliefs about spirits, non-human persons, and other-than-human intelligences functioned as coherent worldviews rather than superstition. Essentially encounters with non-human intelligence are not aberrations but culturally patterned experiences have always known how to contextualize.[289]

Placed alongside Oswald, Lowie is a bridge between one observing nature directly and the other listening carefully to human cultures with them both converging on the same reality the book Interdimensional reveals.

Now looking at Dr. Moses J. Eisenberg, the original owner of all these letters and correspondence, Amy and I discovered that he was an early 20th century physician and psychologist who wrote on abnormal psychology, hallucination and the limits of material explanations for extraordinary human experiences. He treated anomalous perception as data rather than delusion.[290]

So now we have nature (Oswald), culture (Lowie) and mind (Eisenberg) all pointing towards the same unresolved presence humanity keeps encountering under different names.

Dr. James Papez (1893-1958) was a neuroanatomists best known for the Papez circuit, a brain network linking the hippocampus, thalamus and cingulate gyrus, proposing that emotion is no localized but circulatory moving through the brain as an integrated system. His work quietly undermined reductionism by showing that feeling, memory, meaning, and perception are inseparable, which opens the door to anomalous experience being structural, not pathological. If consciousness is a loop, not a point, then contact, intuition, and "other intelligence" may enter through resonance rather than sensation.

Giovanni Schiaparelli (1835-1910) was a pioneering Italian astronomer whose work Die Astronomie im Alten Orient / Astronomy in the Old World traced how ancient civilizations understood the cosmos as animated, intentional, and communicative rather than inert. He showed that early astronomy wasn't just measurement but a relationship. The stars, planets, and phenomena were experienced as intelligences interacting with humanity across myth, ritual and observation. Modern encounters are not new intrusions but the return of an ancient way of seeing the sky as alive and responsive!

And it keeps going, various disciplines are seeing and coming to the same conclusions.

George Miller Sternberg (1838-1915) was a U.S. Army physician, pioneering bacteriologist, and Surgeon General who helped establish modern germ theory while remaining unusually open to the limits of scientific certainty in explaining life and disease.

Working at the edge of the visible and invisible with microbes before they were fully understood, he embodied a moment when science was actively discovering hidden intelligences shaping human reality. Sternberg grounded everything materially and even the most rigorous institutions were forced to accept that unseen agents profoundly influence us. It just reinforces that what may be invisible to us is not imaginary.

Charles Felton Pidgin (1848-1923) was an American historian and writer best known for documenting alternative histories, suppressed traditions, and the persistence of mythic and non-ordinary beliefs beneath official narratives.

He treated folklore, spiritual traditions, and anomalous accounts as historical signals rather than cultural errors, arguing that something meaningful was being remembered imperfectly across time. He was the archivist of the margins. He was the one that showed that what survives in fragments, letters, and forgotten books may be closer to the truth than what institutions chose to preserve.

Another letter was from Henry Harland (1861-1905), an American writer and editor best known as the driving force behind The Yellow Book, the influential literary journal that explored symbolism, mysticism, psychology, and the unseen currents beneath modern life.

Writing under the pseudonym Sidney Luska, Harland was fascinated by liminal states, where art, identity, and hidden forces interact, well before psychology and parapsychology formalized these ideas.[291] He is the literary counterpart to all of these thinkers ahead of their time.

He outlines proof that artists and editors were sensing the same non-human, non-material pressures long before science had language for them. Both of his signatures are on a letter to Dr. Eisenberg.

We couldn't believe the connection of these amazing minds from the past and how it directly lined up with my book at the time I am working on the last page of it when she brought in the letters. We had their signatures, their collective energy in the palm of our hands. We continued to gently look at the names.

Another letter, Paul Rittenhouse, who was a lesser-documented 19th to early 20th century intellectual figure, most often appearing in correspondence, regional publications or private collections rather than major academic histories, which suggests he operated on the edge of formal institutions.

Figures like him functioned as connectors, collectors, commentators and correspondents who circulated ideas between scientists, philosophers, writers and reformers. Non-human intelligence wasn't confined to just the elites of the day, but lived in letters, private networks and personal archives, just like the folder that found its way to me at the right time.

And lastly, Jean-Jacques Rousseau, who was often abbreviated or miswritten as M.J. Rousseau was an enlightenment philosopher who argued that human consciousness, morality and perception are shaped by nature and forces beyond pure rationalism. He believed that civilization dulled an innate sensitivity to deeper truths, positioning intuition and natural awareness as legitimate ways of knowing alongside reason. Non-human or non-material intelligence is something that humanity has lost touch with, not something newly discovered which strengthens the idea of recovery rather than revelation.

As Amy unearthed that old box of letters right when I was finishing the book, it registered as a final nod from the universe. Each figure, Oswald, Lesquereux, Lowie, and the others mirrored the very same themes of intelligence and consciousness I had been exploring.

What mattered was not coincidence, but recognition. These letters surfaced at the precise moment the book reached its end, not as an interruption, but as an affirmation. It delivered evidence that this inquiry has been underway long before me, moving quietly through minds, disciplines, and centuries.

The garage did not offer a discovery; it offered confirmation that the intelligence explored in these pages has always been listening back.

Notes:

1. Jacques Vallee, *Dimensions, A Casebook for Alien Contact* (Chicago: Henry Regnery, 1988), 23.
2. Jacque Vallee, *Dimensions*, 34.
3. John A. Keel, *Operation Trojan Horse* (New York: G.P. Putnam's Sons, 1970), 112.
4. John A. Keel, *The Eighth Tower* (New York: Saturday Review Press, 1975), 89.
5. Jacques Vallee, *Messengers of Deception* (Berkley: And/Or Press, 1979), 18.
6. Colm A. Kelleher and George Knapp, *Hunt for the Skinwalker* (New York: Paraview Pocket Books, 2005), 209–214.
7. John A. Keel, *The Eighth Tower* (New York: Saturday Review Press, 1975), 35-36
8. John Green, *Sasquatch: The Apes Among Us* (Surrey, BC: Hancock House, 1978), 112-115
9. Russell Targ and Harold Puthoff, *Mind Reach* (New York: Delacorte Press, 1977), 7–12; CIA, "Stargate Project," declassified documents.
10. Jacques Vallée, *Dimensions* (New York: Ballantine Books, 1988), 27–29.
11. Nikola Tesla, *My Inventions: The Autobiography of Nikola Tesla,* ed. Ben Johnston (New York: Dover Publications, 1982), 91–92.
12. Albert Einstein, *"Does the Inertia of a Body Depend Upon Its Energy Content?"* Annalen der Physik 18 (1905): 639–641.
13. Albert Einstein and Leopold Infeld, *The Evolution of Physics* (New York: Simon & Schuster, 1938), 221.
14. Brian Greene, *The Elegant Universe* (New York: W. W. Norton, 1999), 95–101.

15. Eric R. Kandel et al., *Principles of Neural Science*, 6th ed. (New York: McGraw-Hill, 2021), 154–160.
16. Rollin McCraty et al., "The Energetic Heart," HeartMath Research Center (2015).
17. Werner Heisenberg, Physics and Philosophy (New York: Harper, 1958), 54–58.
18. Niels Bohr, "The Quantum Postulate and the Recent Development of Atomic Theory," Nature 121 (1928).
19. Brian Greene, *The Fabric of the Cosmos* (New York: Knopf, 2004), 135–142.
20. Joe Dispenza, interview on Impact Theory with Tom Bilyeu, episode aired May 2019.
21. Joe Dispenza, *Becoming Supernatural* (New York: Hay House, 2017), 201–215.
22. Sir John Woodroffe (Arthur Avalon), The Serpent Power (Madras: Ganesh & Co., 1919).
23. David J. Silverman, Faith and Boundaries (Cambridge: Cambridge University Press, 2005).
24. Jacques Vallée, *Passport to Magonia* (Chicago: Henry Regnery, 1969), 20–25.
25. John E. Mack, *Abduction* (New York: Scribner, 1994), 410–420.
26. U.S. Defense Intelligence Agency, Anomalous Acute and Subacute Field Effects (2022).
27. NASA, "Unidentified Anomalous Phenomena Independent Study Team Report," 2023.
28. U.S. House Committee on Oversight and Accountability, "Task Force on the Declassification of Federal Secrets," press release, February 2025.
29. Joe Rogan, The Joe Rogan Experience, interview with Rep. Anna Paulina Luna, August 2025.
30. Brian Greene, *The Elegant Universe* (New York: W. W. Norton, 1999), 187–203.
31. Michio Kaku, *Hyperspace* (New York: Oxford University Press, 1994), 67–89.
32. Hugh Everett III, "Relative State Formulation of Quantum Mechanics," Reviews of Modern Physics 29, no. 3 (1957).

33. Sean Carroll, *Something Deeply Hidden* (New York: Dutton, 2019), 45–78.
34. J. Scudder et al., "First MMS Observations of Electron Diffusion Regions," Nature Physics 8 (2012): 1–7.
35. NASA Goddard Space Flight Center, Magnetospheric Multiscale Mission (MMS) Overview, 2012–present.
36. Corrado Malanga and Filippo Biondi, SAR Tomography Study of the Giza Plateau, press release and preprint, March 15, 2025.
37. Nikola Tesla, *Man's Greatest Achievement*, 1907 writings and lectures.
38. Linda Moulton Howe, Earthfiles, interviews with Douglas Alan Mutschler, 2012–2017
39. Central Intelligence Agency, Analysis and Assessment of Gateway Process, 1983, declassified.
40. Ibid., 25–34.
41. Roger Penrose and Stuart Hameroff, "Consciousness in the Universe: A Review of the Orch OR Theory," Physics of Life Reviews 11, no. 1 (2014): 39–78.
42. Karl H. Pribram, Brain and Perception: Holonomy and Structure in Figural Processing (Hillsdale, NJ: Lawrence Erlbaum, 1991).
43. Joseph L. Kirschvink et al., "Magnetite Biomineralization in the Human Brain," Proceedings of the National Academy of Sciences 89, no. 16 (1992): 7683–7687.
44. Michael A. Persinger, "The Neuropsychiatry of Paranormal Experiences," *Journal of Neuropsychiatry* 12 (2001): 515–524.
45. Johnjoe McFadden, *Life on the Edge: The Coming of Age of Quantum Biology* (New York: Crown, 2014), 201–225.
46. NASA, "*The Electromagnetic Spectrum*," NASA Science Mission Directorate, accessed 2025. 2-4
47. Planck Collaboration, "Planck 2018 Results. VI. Cosmological Parameters," Astronomy & Astrophysics 641 (2020). 6-9
48. Deepasri Prasad and Wilma A. Bainbridge et al., "The Visual Mandela Effect as Evidence for Shared and Specific

False Memories across People," Psychological Science 33, no. 3 (2022). 1971-1988

49. Shawn Ryan, host, "#95 Joe McMoneagle - CIA's Project Stargate," The Shawn Ryan Show podcast audio, February 5 2024 04:06:13-04:07:02
50. Michio Kaku, *Parallel Worlds* (New York: Anchor Books, 2005). 23
51. Sean Carroll, *Something Deeply Hidden* (New York: Dutton, 2019).
52. Bókkon, István. "Biophotons in the Brain: Possible Role in Neural Communication." Progress in Neurobiology 95, no. 4 (2011): 446–456.
53. Rovelli, Carlo. Reality Is Not What It Seems: The Journey to Quantum Gravity. New York: Riverhead Books, 2017, 115–120.
54. Ferri, Francesco et al. "Evidence of Supersolid Light." Nature 620 (2023): 57–62.
55. Reimann, Michael W., et al. "Cliques of Neurons Bound into Cavities Provide a Missing Link between Structure and Function." Frontiers in Computational Neuroscience 11 (2017): 48.
56. Revonsuo, Antti. "The Reinterpretation of Dreams: An Evolutionary Hypothesis." Behavioral and Brain Sciences 23, no. 6 (2000): 877–901.
57. Raduga, Michael. "First Bidirectional Communication Between Dreamers." REMspace Press Release, October 2024.
58. U.S. Defense Intelligence Agency. Stargate Project Archive. Declassified Documents, 1972–1995.
59. Targ, Russell, and Harold E. Puthoff. Mind-Reach: Scientists Look at Psychic Abilities. New York: Delacorte Press, 1977.
60. Allyson Chiu, "The Brains of People Who Say They've Had a UFO Encounter," New York Post, December 12, 2021.
61. Garry P. Nolan, interviews and public lectures on neurological traits observed in UAP experiencers, Stanford University School of Medicine, 2021–2023.

62. Uri Geller, public statements and interviews discussing neurological imaging and reported brain scans following psychic research claims, October 2024.
63. Diane Hennacy Powell, M.D., research as presented in The Telepathy Tapes podcast, 2024.
64. The Telepathy Tapes, official podcast and supporting documentation, 2024.
65. Linxing Preston Jiang et al., "BrainNet: A Multi-Person Brain-to-Brain Interface for Direct Collaboration Between Brains," Scientific Reports 9 (2019): 6115.
66. University of Washington News, "How You and Your Friends Can Play a Video Game Together Using Only Your Minds," July 1, 2019.
67. James Fox, Moment of Contact: New Evidence Revealed (documentary film), 2024.
68. Dr. Italo Venturelli, interview in Moment of Contact: New Evidence Revealed, discussing the 1996 Varginha incident, 2024.
69. Chris Bledsoe, UFO of GOD (New York: Rowman & Littlefield, 2023).
70. Jim Semivan, foreword, and John B. Alexander, introduction, in Chris Bledsoe, UFO of GOD.
71. Jesse Michels, American Alchemy, interview discussing the "indigestible" nature of the UFO truth, 2024.
72. Reporting on NASA's expanded coordination with U.S. intelligence and national security agencies, 2024–2025.
73. José Gabriel Funes, "The Alien Is My Brother," L'Osservatore Romano, May 14, 2008; and statements by Father Richard D'Souza, Vatican Observatory, 2025.
74. Jesse Michels, American Alchemy, interview with Kirsan Ilyumzhinov, 2024.
75. Kirsan Ilyumzhinov, public statements and interviews regarding his 1997 encounter and subsequent discussions with Russian and U.S. political figures.
76. Barbara Marciniak, Bringers of the Dawn: Teachings from the Pleiadians (Rochester, VT: Bear & Company, 1992).

77. Close Encounters of the Third Kind, directed by Steven Spielberg (Columbia Pictures, 1977); U.S. theatrical release November 16, 1977.
78. Jerry R. Ehman, "The Wow! Signal," Big Ear Radio Observatory, Ohio State University, August 15, 1977.
79. "Southern Television Broadcast Interruption," November 26, 1977, United Kingdom.
80. Avi Loeb, "Was the 'Wow! Signal' Emitted from 3I/ATLAS?," Medium, October 2025.
81. Steven Spielberg, Disclosure Day (film trailer and promotional materials), released December 2025.
82. NASA and European Space Agency public statements regarding interstellar object 3I/ATLAS and planetary defense observation exercises, December 2025.
83. Uri Geller, Facebook post, December 19, 2025.
84. Dena J. Dincauze, The Archaeology of New England (New York: Academic Press, 1990), 213–216; William Simmons, Spirit of the New England Tribes (Hanover, NH: University Press of New England, 1986), 98–102.
85. Jerome Clark, The UFO Book: Encyclopedia of the Extraterrestrial (Detroit: Visible Ink Press, 1998), 36–38; David M. Jacobs, The UFO Controversy in America (Bloomington: Indiana University Press, 1975), 52–54.
86. MIT Lincoln Laboratory, History of the Lincoln Laboratory Radar and Space Surveillance Programs (Lexington, MA: MIT Lincoln Laboratory Publications, 2003), 1–12.
87. U.S. Army Natick Soldier Systems Center, A History of the U.S. Army Soldier Systems Command (Natick, MA: Department of the Army, 2007), 5–18.85.
88. George Noory, interview on Beyond Belief, Gaia TV, Season 14 episode featuring Ronny LeBlanc (2023).
89. Charles Fort, The Book of the Damned (New York: Boni and Liveright, 1919), 35–37; Charles Fort, Lo! (New York: Claude Kendall, 1931), 281–283.
90. Karl T. Pflock, Roswell: Inconvenient Facts and the Will to Believe (Amherst, NY: Prometheus Books, 2001), 1–25;

Don Berliner and Stanton T. Friedman, Crash at Corona (New York: Paraview, 1992), 3–18.

91. John L. Schuessler, UFOs: Generals, Pilots, and Government Officials Go on the Record (New York: MJF Books, 1997), 170–182; Kevin D. Randle, UFO Casebook (New York: HarperCollins, 1989), 222–229.
92. Kenneth Arnold, "I Did See the Flying Discs," Fate Magazine 1, no. 1 (1948): 5–11; Jerome Clark, The UFO Book: Encyclopedia of the Extraterrestrial (Detroit: Visible Ink Press, 1998), 15–18.
93. Federal Aviation Administration, "Temporary Flight Restriction: Bedminster, NJ," FAA Notice, November 22, 2024.
94. Federal Aviation Administration, "Temporary Flight Restriction: Picatinny Arsenal," FAA Notice, November 25, 2024.
95. Federal Bureau of Investigation, Newark Field Office, public statement on coordinated investigation into reported drone activity, December 2024.
96. Office of Governor Phil Murphy, Press Briefings on unidentified drone activity, December 5 and December 9, 2024.
97. U.S. Department of Defense, Pentagon Press Briefing, December 11, 2024.
98. Kathy Hochul, Governor of New York, Press Statement on federal drone-detection deployment, December 14, 2024.
99. John Kirby, National Security Council, Press Briefing, December 12, 2024.
100. U.S. Air Force, Langley Air Force Base incident reports on repeated drone incursions, December 2023.
101. U.S. Navy, Unidentified Aerial Phenomena reports involving USS Kidd, USS Rafael Peralta, and USS Russell, 2019.
102. NewsNation, reporting by Rick McHugh on Ocean County drone sightings and law-enforcement encounters, December 2024.

103. Canadian Department of National Defence, Statement on unidentified aerial object shot down over Yukon Territory, February 11, 2023.
104. North American Aerospace Defense Command (NORAD), Press Briefings on high-altitude object detections, February 2023.
105. North American Aerospace Defense Command (NORAD), Timeline of High-Altitude Object Detections, January–February 2023.
106. Beatriz Villarroel et al., "Anomalous Transients in Historical Astronomical Plates," Acta Astronautica (under peer review), Nordic Institute for Theoretical Physics, 2024.
107. Beatriz Villarroel et al., analysis of Palomar Observatory photographic plates (1942–1957).
108. Beatriz Villarroel et al., "The Vanishing and Appearing Sources during the Cold War," Nordic Institute for Theoretical Physics preprint archive, 2023–2024.
109. Jacques Vallée, The Invisible College (New York: E. P. Dutton, 1975), 15–23.
110. J. Allen Hynek, The UFO Experience (Chicago: Henry Regnery Company, 1972), 167–176.
111. U.S. Department of Defense, "Preliminary Assessment: Unidentified Aerial Phenomena," Office of the Director of National Intelligence, June 2021.
112. Project Blue Book Special Report No. 14, U.S. Air Force, 1955.
113. John A. Keel, UFOs: Operation Trojan Horse (New York: G. P. Putnam's Sons, 1970), 296–304.
114. Jewitt, David, et al. "Interstellar Interlopers: 1I/'Oumuamua and 2I/Borisov." Annual Review of Astronomy and Astrophysics 61 (2023): 441–483.
115. Laughlin, Gregory, and Konstantin Batygin. "The Dynamics of Interstellar Objects in the Solar System." Astrophysical Journal Letters 880, no. 2 (2019): L15-19, esp. L16-18

116. Asteroid Terrestrial-impact Last Alert System (ATLAS). "ATLAS Survey and Detection of Interstellar Objects." University of Hawaiʻi Institute for Astronomy, 2025.
117. Loeb, Avi. "Could Interstellar Objects Be Technological Artifacts?" Scientific American, February 2023. 32-37
118. Loeb, Avi. Extraterrestrial: The First Sign of Intelligent Life Beyond Earth. Boston: Houghton Mifflin Harcourt, 2021. 87-112
119. Gray, Robert H. "The Wow! Signal: A Forty-Year Search for the Source." Journal of the Washington Academy of Sciences 103, no. 3 (2017). 46-55
120. National Aeronautics and Space Administration. "Mars Reconnaissance Orbiter Mission Overview." NASA JPL, updated 2024.
121. Executive Order 14347. "Restoring the United States Department of War." Signed September 5, 2025; Federal Register, September 10, 2025. 43893-43895
122. End of the clarifying sentence noting that the Department of War originally became the Department of Defense in 1947.
123. National Aeronautics and Space Administration. "NASA to Streamline Agency Social Media Accounts." NASA Press Release, July 11, 2025.
124. Congressional Research Service. "Shutdown of the Federal Government: Causes, Processes, and Effects." Updated 2024. pp.6-9
125. Pope, Nick. Encountering the Unexplained: UFOs and the Ministry of Defence. London: Simon & Schuster, 2014. 223-226
126. Jackson, Patrick. The Sphere Network. Independently published, February 15, 2025. 14-52, esp. 31-44
127. Clark, Jerome. Unexplained! Strange Sightings, Incredible Occurrences, and Puzzling Physical Phenomena. Detroit: Visible Ink Press, 2016.312-329, esp. 318-322
128. Betz, Antoine. "The Betz Sphere Incident." Journal of UFO Studies 2, no. 1 (1979). 9-21, esp. 11-15

129. Mossbridge, Julia. Public commentary on anomalous artifacts and perception. Institute of Noetic Sciences, 2024.
130. Morcillo, P. "A Unified Framework for the Buga Sphere: Quantitative Validation of a Negative-Mass Model Governed by Topo-Temporal Physics." Preprint manuscript, 2025. 1-18, esp. 7-14
131. Greer, Steven M. Unacknowledged. Los Angeles: Paradigm Research Group, 2017. 287-304, esp. 296-300
132. Coulthart, Ross. In Plain Sight. Melbourne: Hardie Grant Books, 2021. 182-191
133. W. Y. Evans-Wentz, The Fairy-Faith in Celtic Countries (London: Oxford University Press, 1911), 31–33, 284–286.
134. William S. Simmons, Spirit of the New England Tribes: Indian History and Folklore, 1620–1984 (Hanover, NH: University Press of New England, 1986), 217–219; Kathleen J. Bragdon, Native People of Southern New England, 1500–1650 (Norman: University of Oklahoma Press, 1996), 243–245; Charles G. Leland, The Algonquin Legends of New England (Boston: Houghton, Mifflin and Company, 1884), 89–94.
135. Kathleen J. Bragdon, Native People of Southern New England, 243–245.
136. Charles G. Leland, The Algonquin Legends of New England, 89–94; John A. Keel, The Eighth Tower (New York: Saturday Review Press, 1975), 35–38.
137. Loren Coleman, Mysterious America (New York: Paraview Pocket Books, 1983), 269–276; Walter J. Nichols, The Bridgewater Triangle (Charleston, SC: History Press, 2013), 41–44.
138. Jacques Vallée, Dimensions: A Casebook of Alien Contact (New York: Ballantine Books, 1988), 17–21, 153–156.
139. History Channel, MonsterQuest, "Ohio Grassman," Season 3, Episode 10, aired January 6, 2010; Animal Planet, Finding Bigfoot, "Ohio Grassman," Season 4, Episode 2, aired January 19, 2013.
140. John A. Keel, The Eighth Tower, 189–192.

141. Jacques Vallée, Dimensions, 153–156; Jeffrey J. Kripal, Authors of the Impossible (Chicago: University of Chicago Press, 2010), 201–204.
142. John A. Keel, The Eighth Tower, 189–192.
143. Jacques Vallée, Dimensions, 153–156.
144. Figures 11.1-11.2 reproduced with permission, eyewitness and author photographs, Leominster, Massachusetts, August - September 2023
145. Luis Elizondo, interview with NewsNation, statements on physiological and neurological effects observed in military personnel following UAP encounters, 2022–2023.
146. Nick Pope, Encounter in Rendlesham Forest: The Inside Story of the World's Best-Documented UFO Incident (London: John Blake, 2014), 173-176; U.S. Department of Veterans Affairs, disability ruling for Airman John Burroughs related to Rendlesham Forest incident, 2015.
147. Robert E. Bartholomew and Simon Wessely, "Protean nature of mass sociogenic illness," British Journal of Psychiatry 185 (2004): 300-306, esp. 302-304
148. Annie Jacobsen, Phenomena (New York: Little, Brown, 2023), 287–293.
149. National Academies of Sciences, Engineering, and Medicine, An Assessment of Illness in U.S. Government Employees and Their Families at Overseas Embassies (Washington, DC: National Academies Press, 2020), 5–8, 77–92.
150. David Grusch interview with NewsNation, June 2023. Ralph Blumenthal and Leslie Kean, "UFO Whistleblower Says Government Has Nonhuman Craft," New York Times, June 5, 2023.
151. Christopher K. Daugherty et al., "Functional brain imaging of affected U.S. diplomats," JAMA 319, no. 11 (2018): 1125–1135, esp. 1128–1132.
152. Leslie Kean, Ralph Blumenthal, and Garry Nolan, "Anomalous Acute and Subacute Field Effects on Human Biological Tissues," The Debrief, May 2021.

153. Garry Nolan interview, The Stanford University School of Medicine Colloquium, 2022.
154. Luis Elizondo, testimony summarized in Politico and The Hill, 2022 congressional briefings.
155. Office of the Inspector General of the Intelligence Community, statement on David Grusch whistleblower complaint, July 2023.
156. U.S. Army Intelligence and Security Command, Project Stargate Final Report, 1995, declassified.3–7; esp. 79–85.
157. Michael Persinger, Neuropsychological Bases of God Beliefs (New York: Praeger, 1987), 41–58
158. Eugene d'Aquili and Andrew Newberg, The Mystical Mind (Minneapolis: Fortress Press, 1999), 89–117.
159. Edward J. Ruppelt, The Report on Unidentified Flying Objects (Garden City, NY: Doubleday, 1956), 195–201.
160. Department of Defense, "UAP Encounters by Naval Aviators," briefing materials, 2019–2021.
161. Nikola Tesla, My Inventions (New York: Electrical Experimenter Publishing, 1919), 91–95.
162. William L. Moore and Charles Berlitz, The Philadelphia Experiment: Project Invisibility (New York: Pocket Books, 1979), 21–38, 187.
163. Morris K. Jessup, The Case for the UFO (New York: Citadel Press, 1955), Appendix materials.
164. Charles Berlitz, The Bermuda Triangle (Garden City, NY: Doubleday, 1974), cross-referenced interviews.
165. St. Albans Weekly Messenger, "Wild Man of the Woods," February 8, 1855.
166. Waterville Sentinel, "A Wild Man Killed in the Maine Woods," 1886.
167. Personal field investigation conducted by the author with the Maine Bigfoot Foundation, August–September 2024.
168. John E. Mack, Passport to the Cosmos (New York: Crown Publishers, 1999), 215–222.
169. Jacques Vallée, Dimensions (Chicago: Henry Regnery, 1988), 171–178.

170. Grover S. Krantz, Big Footprints: A Scientific Inquiry into the Reality of Sasquatch (Boulder, CO: Johnson Books, 1992), 79–82.
171. Christopher Packard, Mythical Creatures of Maine (Charleston, SC: The History Press, 2014), 42–47.
172. Jeffrey J. Kripal, The Flip: Epiphanies of Mind and the Future of Knowledge (New York: Bellevue Literary Press, 2019), 63–70.
173. Figure 13.1, Author's photograph with 20-inch cast, Secret Location, Maine, August 12, 2024
174. Jacques Vallée, Passport to Magonia: From Folklore to Flying Saucers (Chicago: Henry Regnery, 1969), 15–38.
175. Joshua Cutchin, Ecology of Souls (San Antonio: Anomalist Books, 2014), 133–159.
176. Chris Aubeck and Martin Shough, Return to Magonia (Exeter: Triarchy Press, 2015), 201–230.
177. Wilson Duff, The Indian History of British Columbia (Victoria: British Columbia Provincial Museum, 1964), 62–65.
178. Thomas Buckley, Standing Ground: Yurok Indian Spirituality (Berkeley: University of California Press, 2002), 214–217.
179. Alfred L. Kroeber, Handbook of the Indians of California (Washington, DC: Smithsonian Institution, 1925), 110–112.
180. David Thompson, David Thompson's Narrative of His Explorations in Western America, 1784–1812 (Toronto: Champlain Society, 1916), 92–93.
181. Ronald Hutton, The Pagan Religions of the Ancient British Isles (Oxford: Blackwell, 1991), 217–220.
182. Paul Devereux, Places of Power (London: Blandford Press, 1990), 67–88.
183. J.W. Burns, "Introducing B.C.'s Hairy Giants," Maclean's Magazine, April 1, 1929.
184. Diamond Jenness, The Indians of Canada (Ottawa: National Museum of Canada, 1932), 181–183.
185. Lady Raglan (Julia Somerset), "The Green Man in Church Architecture," Folklore 50, no. 1 (1939): 45–57.

186. Kathleen Basford, The Green Man (Cambridge: D.S. Brewer, 1978), 18–41.
187. Miranda Green, Symbol and Image in Celtic Religious Art (London: Routledge, 1989), 102–109.
188. Katherine Briggs, An Encyclopedia of Fairies (New York: Pantheon, 1976), 180–184.
189. James Frazer, The Golden Bough, abridged ed. (New York: Macmillan, 1922), 92–99; William Haviland and Marjory Power, The Original Vermonters (Hanover, NH: University Press of New England, 1994), 271–273.
190. Stanley Krippner and David Feinstein, Personal Mythology (Los Angeles: Tarcher, 1992), 201–205.
191. Ronald Hutton, The Triumph of the Moon (Oxford: Oxford University Press, 1999), 312–315.
192. John E. Mack, Abduction: Human Encounters with Aliens (New York: Scribner, 1994), 49–72.
193. Edith Turner, Experiencing Ritual (Philadelphia: University of Pennsylvania Press, 1992), 123–129.
194. Carl Jung, Synchronicity: An Acausal Connecting Principle (Princeton: Princeton University Press, 1973), 84–96.
195. Joseph Epes Brown, The Sacred Pipe (Norman: University of Oklahoma Press, 1953), 88–92.
196. Linda S. Cordell and Maxine McBrinn, Archaeology of the Southwest (Walnut Creek, CA: Left Coast Press, 2012), 289–291.
197. Mircea Eliade, The Sacred and the Profane: The Nature of Religion, trans. Willard R. Trask (New York: Harcourt, Brace & World, 1959), 10–12.
198. Paul Devereux, Earth Lights Revelation (London: Gothic Image Publications, 1989), 21–35.; Michael Persinger, "Geophysical Variables and Behavior," Perceptual and Motor Skills 51, no. 1 (1980): 3–16.
199. Jacques Vallée, Passport to Magonia: From Folklore to Flying Saucers (Chicago: Henry Regnery, 1969), 9.
200. Colm A. Kelleher and George Knapp, Hunt for the Skinwalker: Science Confronts the Unexplained at a Remote

Ranch in Utah (New York: Paraview Pocket Books, 2005), 25–41.
201. Kelleher and Knapp, Hunt for the Skinwalker, 293–300.
202. Larry Warren and Peter Robbins, Left at East Gate (Charleston, SC: History Press, 2014), 62–85; Charles Halt, “Unexplained Lights,” memorandum, RAF Bentwaters, January 13, 1981.
203. Patrick Jackson, Quantum Paranormal (self-published, 2019), 112–130.
204. Hildburgh, W. L., “The Wild Man in Medieval Art,” Folklore 47, no. 2 (1936): 134–151; Adrienne Mayor, The First Fossil Hunters (Princeton: Princeton University Press, 2000), 258–263.
205. Edith Turner, Communitas: The Anthropology of Collective Joy (New York: Palgrave Macmillan, 2012), 89.
206. James George Frazer, The Golden Bough (New York: Macmillan, 1922), 59–63.
207. John Bierhorst, The Mythology of North America (New York: William Morrow, 1985), 87–90.
208. John E. Mack, Passport to the Cosmos: Human Transformation and Alien Encounters (New York: Crown, 1999), 4–6.
209. Figure 15.1a and Figure 15.1b, photographs taken by Kayleigh Baysinger during the 2025 Get Haunted Trifecta event at Fort Duffield, West Point, Kentucky; Figure 15.1b is a cropped enlargement of Figure 15.1a, included for detail clarity.
210. Figure 15.2, photograph taken by Kayleigh Baysinger at Fort Duffield, West Point, Kentucky, 2025, documenting the location associated with the reported cloaking encounter. Grey alien looking being captured on photograph.
211. Figure 15.3, artistic rendering by Bill Bryan, 2025, illustrating the cloaked humanoid figure he reports observing at Fort Duffield, West Point, Kentucky.
212. Davis, Erik. TechGnosis: Myth, Magic, and Mysticism in the Age of Information. New York: Harmony Books, 1998, 174–177.

213. Keel, John A. The Eighth Tower. New York: Saturday Review Press, 1975, 37–41.
214. Jung, Carl G. Synchronicity: An Acausal Connecting Principle. Princeton: Princeton University Press, 1973, 1–6.
215. Vallee, Jacques. Dimensions. New York: Ballantine Books, 1988, 91–94.
216. Vallee, Jacques. Passport to Magonia. Chicago: Henry Regnery, 1969, 215–218.
217. Mack, John E. Abduction. New York: Scribner, 1994, 49–52.
218. Keel, John A. The Mothman Prophecies. New York: Tor Books, 2002, 88–90.
219. Persinger, Michael A. Neuropsychological Bases of God Beliefs. New York: Praeger, 1987, 67–70.
220. Vallee, Jacques. Confrontations. New York: Ballantine Books, 1990, 112–115.
221. Ring, Kenneth. The Omega Project. New York: William Morrow, 1992, 101–104.
222. Keel, John A. Strange Creatures from Time and Space. New York: Fawcett Gold Medal, 1970, 146–149.
223. Bender, Hans. "Psi Phenomena and Heredity." Journal of Parapsychology 30, no. 3 (1966): 201–205.
224. Haines, Richard F. UFO Phenomena and the Behavioral Scientist. Lanham: Scarecrow Press, 2003, 132–135.
225. Hopkins, Budd. Intruders. New York: Random House, 1987, 58–62.
226. Office of the Director of National Intelligence. Preliminary Assessment: Unidentified Aerial Phenomena. Washington, DC, 2021, 6–7.
227. Levine, George. UFOs and Alien Abduction Phenomena. New York: Berkley, 1998, 143–146.
228. Mack, John E. Abduction. New York: Scribner, 1994, 391–394.
229. Andreasson, Betty. The Andreasson Affair. New York: Prentice-Hall, 1979, 23–26.
230. Ring, Kenneth. Heading Toward Omega. New York: William Morrow, 1984, 214–217.

231. Eliade, Mircea. Shamanism: Archaic Techniques of Ecstasy. Princeton: Princeton University Press, 1964, 505–508.
232. LeBlanc, Ronny. Monsterland. Self-published, 2017, 112–115
233. Bindernagel, John. North America's Great Ape. Blaine, WA: Hancock House, 1998, 187–190.
234. Todd, Leslie. "Infrasound and Large Mammals." Journal of Acoustical Ecology 12, no. 2 (2006): 44–47.
235. Devereux, Paul. Earth Lights. London: Turnstone Press, 1982, 119–122.
236. Morehead, Ron, and Al Berry. The Sierra Sounds. Audio recordings, 1971–1974.
237. Green, Grover Krantz. Big Footprints. Boulder: Johnson Books, 1992, 151–154.
238. Mack, John E. Passport to the Cosmos. New York: Crown, 1999, 33–36.
239. James, William. The Varieties of Religious Experience. New York: Longmans, Green & Co., 1902, 379–381.
240. Vallee, Jacques. The Invisible College. New York: Dutton, 1975, 206–210.
241. Salvatore Cezar Pais, interview in "Salvatore Pais, a Brilliant DoD Engineer Discussing the Plasma & Consciousness Connection," public lecture/interview, circulated online, c. 2020–2021.
242. Salvatore Cezar Pais, U.S. Patent No. 10,135,366, Craft Using an Inertial Mass Reduction Device, filed 2016, granted 2018; and related patents assigned to U.S. Navy.
243. Francis F. Chen, Introduction to Plasma Physics and Controlled Fusion, 2nd ed. (New York: Springer, 1984), 1–5.
244. Hannes Alfvén, Cosmic Plasma (Dordrecht: Reidel, 1981), 3–12.
245. Irving Langmuir, "Oscillations in Ionized Gases," Proceedings of the National Academy of Sciences 14, no. 8 (1928): 627–637.
246. Salvatore Cezar Pais, "Plasma Compression Fusion Device," U.S. Patent No. 10,058,114 (2018).

247. L. Liu et al., "Remote Sensing of Equatorial Plasma Bubbles Using LARID Radar," Chinese Journal of Space Science (2024).
248. National Aeronautics and Space Administration (NASA), "Equatorial Plasma Irregularities and Space Weather," NASA Science Brief, updated 2023.
249. U.S. House of Representatives, Hearing: Restoring Public Trust Through UAP Transparency and Whistleblower Protection, testimony of Dylan Borland, September 9, 2025.
250. Jacques Vallée and Eric Davis, "Incommensurability, Orthodoxy and the Physics of High Strangeness," Journal of Scientific Exploration 17, no. 1 (2003): 37–71.
251. Carl Jung, Memories, Dreams, Reflections, trans. Richard and Clara Winston (New York: Vintage, 1989), 356.
252. The Acts of the Apostles, Acts 2:1–6, Greek New Testament (Nestle-Aland 28th ed.).
253. Bart D. Ehrman, Jesus: Apocalyptic Prophet of the New Millennium (New York: Oxford University Press, 1999), 194–197.
254. Mircea Eliade, Shamanism: Archaic Techniques of Ecstasy (Princeton: Princeton University Press, 1964), 38–45.
255. Gershom Scholem, Major Trends in Jewish Mysticism (New York: Schocken, 1974), 20–27.
256. The Book of Enoch (1 Enoch), trans. R.H. Charles (Oxford: Clarendon Press, 1893), 14–16.
257. Ignatius of Antioch, Letter to the Ephesians, early 2nd century.
258. John Chrysostom, Homilies on Matthew, Homily 6.
259. William James, The Varieties of Religious Experience (New York: Longmans, Green & Co., 1902), 388.
260. Stanislav Grof, The Holotropic Mind (New York: HarperCollins, 1992), 27–35.
261. Pierre Teilhard de Chardin, The Phenomenon of Man (New York: Harper & Row, 1959), 261.
262. Jamie Lewis and Andrew Bartlett, Bigfooters and Scientific Inquiry (Cardiff: Cardiff University Press, forthcoming).

263. Dan Farah, *The Age of Disclosure*, documentary film, 2024.
264. David Paulides, *American Sasquatch*, documentary film, 2025.
265. James Fox, *Moment of Contact: New Revelations of Alien Contact*, documentary film (forthcoming update).
266. Steven Spielberg, *Disclosure Day* a UFO/UAP disclosure film, announced 2024; Times Square promotional billboard, 2025.
267. Lewis and Bartlett, Bigfooters and Scientific Inquiry.
268. Harry Collins, Changing Order: Replication and Induction in Scientific Practice (Chicago: University of Chicago Press, 1985), esp. discussion of "experimenter's regress."
269. Paul Feyerabend, Against Method (London: Verso, 1975).
270. Charles Fort, The Book of the Damned (New York: Boni and Liveright, 1919).
271. Thomas S. Kuhn, The Structure of Scientific Revolutions, 2nd ed. (Chicago: University of Chicago Press, 1970).
272. Paul G. Bahn and Jean Vertut, Journey Through the Ice Age (Berkeley: University of California Press, 1997), on historical resistance to anomalous discoveries (meteorites, gorillas, platypus).
273. Lewis and Bartlett, Bigfooters and Scientific Inquiry.
274. John E. Mack, Abduction: Human Encounters with Aliens (New York: Scribner, 1994).
275. Stanislav Grof, The Holotropic Mind (New York: HarperCollins, 1992).
276. Jerome Clark, Unexplained! Strange Sightings, Incredible Occurrences, and Puzzling Physical Phenomena (Detroit: Visible Ink Press, 1993).
277. Mircea Eliade, Shamanism: Archaic Techniques of Ecstasy (Princeton: Princeton University Press, 1964).
278. National Institute of Standards and Technology, overview of electromagnetic effects and infrasound phenomena in environmental encounters.

279. Thomas Young, "On the Theory of Light and Colours," Philosophical Transactions of the Royal Society of London 92 (1802).
280. Central Intelligence Agency, "Analysis and Assessment of Gateway Process," declassified document, 1983 (released 2003).
281. Max Tegmark, Our Mathematical Universe (New York: Knopf, 2014).
282. Jacques Vallée, The Invisible College (New York: Dutton, 1975).
283. Jacques Vallée, Passport to Magonia (Chicago: Henry Regnery, 1969).
284. David J. Bohm, Wholeness and the Implicate Order (London: Routledge, 1980).
285. Buddha, Dhammapada, trans. Eknath Easwaran (Tomales, CA: Nilgiri Press, 2007).
286. Felix L. Oswald, The Secret of the East, or The Higher Knowledge (London: Kegan Paul, Trench & Co., 1881), on naturalistic pantheism and non-material intelligence.
287. Felix L. Oswald, Days and Nights in the Tropics (New York: D. Appleton and Company, 1885), esp. chapters on nature, consciousness, and the evolutionary mind.
288. Leo Lesquereux, Manual of the Mosses of North America (Cambridge, MA: Harvard University Press, 1884); see also Louis Agassiz correspondence referencing Lesquereux's paleobotanical work in Letters and Recollections of Alexander Agassiz (Boston: Houghton Mifflin, 1913).
289. Robert H. Lowie, Primitive Religion (New York: Boni and Liveright, 1924), contextualizing non-human intelligence within coherent cultural systems. (Supports Eisenberg/Lowie bridge)
290. Moses J. Eisenberg, "Abnormal Psychic Phenomena and the Limits of Material Explanation," Journal of Abnormal Psychology 14, no. 3 (1919): 145–162.
291. Henry Harland (Sidney Luska), selected editorials and correspondence in The Yellow Book, vols. 1–13 (London: Elkin

Mathews & John Lane, 1894–1897), esp. essays on symbolism, liminality, and unseen forces in modern life.

APPENDIX

Patterns, Terms, and Frameworks Referenced in This Book

A. Key Concepts

Interdimensional Hypothesis · Consciousness as a Non-Local Field · The Hitchhiker Effect · Liminal / Window Areas · Convergence Events · Observer Participation · Plasma-Based Intelligence · Electromagnetic Sensitivity · Portals & Thin Places

B. Phenomena Discussed

UAP / UFO sightings · Orbs & anomalous lights · Sasquatch / Cryptid encounters · Telepathy & mindspeak · Time distortion · Cloaking phenomena · Poltergeist-like activity · Synchronicity · Near-death and altered-state experiences

C. Disciplines Integrated

Theoretical Physics · Neuroscience · Consciousness Studies · Anthropology & Folklore · Intelligence & Defense Research · Archaeology · Parapsychology · Experiential Field Investigation

D. Methodological Note

This work synthesizes peer-reviewed research, declassified intelligence documents, historical sources, and firsthand experiential data. The goal is not to assert final answers, but to map repeating patterns across disciplines where conventional frameworks fail to overlap.

BIBLIOGRAPHY (Selected)

Bohm, David. Wholeness and the Implicate Order.

Dispenza, Joe. Becoming Supernatural.

Hameroff, Stuart, and Roger Penrose. "Orchestrated Objective Reduction."

Keel, John A. The Eighth Tower; Operation Trojan Horse.

Mack, John E. Abduction.

Pribram, Karl. Languages of the Brain.

Tesla, Nikola. Selected Writings and Lectures.

Vallée, Jacques. Passport to Magonia; Dimensions; Revelations.

U.S. Central Intelligence Agency. Analysis and Assessment of the Gateway Process.

NASA & University of Iowa. Magnetospheric Portal Studies.

Congressional Oversight Hearings and Declassified UAP Reports (2023–2025).

About the Author

Ronny LeBlanc is an author, researcher, and television personality best known for his work exploring consciousness, cryptids, and high-strangeness phenomena at the intersection of science, folklore, and human experience.

He is the author of the *Monsterland* books, which document regional patterns of anomalous activity in New England, and his work on Interdimensional expands that foundation into a broader framework linking UFOs, non-human intelligence, and consciousness itself.

LeBlanc has starred in Discovery Channel's *Expedition Bigfoot* and is one of the analysts for Travel Channel's *Paranormal Caught on Camera*, among other programs, and continues to lecture publicly on emerging patterns in UAP research, environmental intelligence, and the evolving nature of human perception.

Printed in Dunstable, United Kingdom

77565868R00170